Advanced Topics in Discrete Optimization

The Author

Er. Shruti Singh did M.Tech (Computer Science and Engineering) from Shri Ramswaroop Memorial University (SRMU) Lucknow and B.Tech (Computer Science and Engineering) from Gautam Buddh Technical University (GBTU), Lucknow. She worked as an Assistant Professor at Himalayan Institute of Technology and Management, Lucknow. She has published number of research papers in national and international journals of repute and also presented papers in several conferences. She has received various honours and awards in the field of computer science and engineering.

Advanced Topics in Discrete Optimization

Author
Shruti Singh

Kruger Brentt
Publishers

2020

Kruger Brentt Publishers UK. LTD.
Company Number 9728962

Regd. Office: 68 St Margarets Road, Edgware, Middlesex HA8 9UU

Disclaimer:

Every possible effort has been made to ensure that the information contained in this book is accurate at the time of going to press, and the publisher and author cannot accept responsibility for any errors or omissions, however caused. No responsibility for loss or damage occasioned to any person acting, or refraining from action, as a result of the material in this publication can be accepted by the editor, the publisher or the author. The Publisher is not associated with any product or vendor mentioned in the book. The contents of this work are intended to further general scientific research, understanding and discussion only. Readers should consult with a specialist where appropriate.

Every effort has been made to trace the owners of copyright material used in this book, if any. The author and the publisher will be grateful for any omission brought to their notice for acknowledgement in the future editions of the book.

Library of Congress Cataloging-in-Publication Data

ISBN 978-1-78715-026-3 (Hardbound)

For information on all our publications visit our website at http://krugerbrentt.com/

Preface

Optimization is the mathematical problem of finding a decision to achieve the best possible outcome while satisfying certain restrictions. Linear programs, conic linear programs and discrete optimization problems arise in a myriad of applications: electricity markets, airlines, logistics, public transport, international shipping, mining, finance, engineering, and data science. This course will provide an introduction to the basic mathematical theory, modelling techniques, computational methods and selected applications of linear, conic and discrete optimization. Linear programming is a method to achieve the best outcome (such as maximum profit or lowest cost) in a mathematical model whose requirements are represented by linear relationships. Linear programming is a special case of mathematical programming.

More formally, linear programming is a technique for the optimization of a linear objective function, subject to linear equality and linear inequality constraints. Its feasible region is a convex polytope, which is a set defined as the intersection of finitely many half spaces, each of which is defined by a linear inequality. Its objective function is a real-valued affine function defined on this polyhedron. A linear programming algorithm finds a point in the polyhedron where this function has the smallest value if such a point exists.

The present book entitled ***Advanced Topics in Discrete Mathematics*** contains six chapters covering all related disciplines. These chapters include: modern approaches to combinatorial optimization, linear programming, polyhedral methods, approximation algorithms, integer programming and constraint programming.

Constructive suggestions of readers especially the teaching faculty and students are invited for further improvement of the book. Their esteemed suggestions will be duly acknowledged in the future editions.

Er. Shruti Singh

Contents

UNIT 1

Modern Approaches to Combinatorial Optimization

1.1 Introduction

Almost all of the results surveyed here owe a debt of one sort or another to the classic theorem of Beardwood, Halton, and Hammersley that lays out the basic behavior of the length of the shortest tour through a random sample from a general distribution in $\mathbf{R}^d$:

Theorem 1.1 If X_i, $1 \leq o \leq \infty$ are independently and identically distributed random variables with bounded support in $\mathbf{R}^d$, then the length L_n under the usual Euclidean metric of the shortest path through the points $\{X_1, X_2, \ldots, X_n\}$ satisfies

$$L_n / n^{(d-1)/d} \to \beta_{TSP,d} \int_{R^d} f(x)^{(d-1)/d} \, dx \text{ almost surely.}$$

Here, $f(x)$ is the density of the absolutely continuous part of the distribution of the X_i, and $\beta_{TSP, d}$ is a positive constant that depends on d but not on the distribution of the X_i.

The Beardwood, Halton, and Hammersley (BHH) theorem has led to developments of several different types. The first developments that we review concern generalizations of the BHH theorem that aim to provide insight into larger classes of processes. It turns out that the essential geometric features of the TSP are present in many of the problems that have been investigated in the theory of combinatorial optimization, and there is also a close connection between these properties and the notions of subadditivity. These sub-additivity-driven generalizations of the BHH are addressed primarily in 1.2.

The second type of problem considered looks at the BHH theorem from the perspective of precision, rather than generality. It turns out that L_n, the length of the shortest tour through an n-sample, is highly concentrated about its mean. The main result of this development, the theorem of Rhee and Talagrand, is spelled out in 1.3. It tells us that the tails of the TSP functional L_n decay as rapidly as those of the normal distribution. This remarkable result has evolved through several intermediate stages, each of which offered new tools for the analysis of functionals of finite point sets in the plane.

Section 1.4 focus explicitly on the theory of the TSP, and, though we start to strain the notion of a Euclidean problem, the development casts light on problems like those just described. The main purpose of 1.4 is to introduce the connection between Euclidean optimization problems and self-similar sets of fractional dimension; the principal result reviewed is a theorem by S. Lalley on the TSP. This is an intriguing result that suggests a rich connection between the rapidly developing theory of sets of fractals and many classical problems.

In 1.5 and 1.6 the set of problems studied is widened. We first consider the theory of the minimal spanning tree (MST) and review how one can construct an infinite analog to the MST that leads to answers of a number of classical questions. One of the best reasons for introducing any new mathematical structure is that it simplifies and strengthens earlier results, and this is exactly what occurs in this development of an "infinite MST." The new structure permits one to make many deductions that are apparently difficult (or perhaps impossible) to address without such a tool.

The 1.6 gives a brief survey of some of the new developments in the theory of matching. This subject began with the theorem of Ajtai, Komlós, and Tusnády on the two-sample matching problem, but it has developed rapidly because of its rich connections to such topics as bin packing, scheduling, and the theory of empirical processes. There are many engaging special results in this active area, and it is not possible to offer much more than a taste. We particularly want to call attention to a new theory of majorizing measure initiated by M. Talagrand. This theory unifies and deepens many earlier results and also suggests many new concrete problems.

1.2 Connection to Algorithms

Before going to the survey just outlined, there is one top-level observation about the BHH theorem that should be made. It is likely that this remarkable 1959 result would have remained relatively undeveloped if it had not been for the striking use of the BHH theorem in the polynomial-time probabilistic TSP algorithm of Karp. The TSP was among the first of the traditional problems of geometric optimization to be proved to be NP-hard, and it is also a problem of genuine practical interest, so it is not surprising that considerable excitement was generated when Karp showed that the TSP is perfectly tractable under the plausible stochastic assumption that the sites to be visited by the tour can be modeled as a random uniform sample. In this context, Karp showed that, given any $\epsilon > 0$, there is a polynomial time algorithm that produces a tour of length no more than $(1 + \epsilon)$ times the optimal tour length with probability that converges to one as the size of the problem is increased. This discovery launched an important branch of the field of probabilistic algorithms. Moreover, it generated powerful interest in the BHH theorem, its generalizations, and sharpenings that are at the center of this survey.

1.3 Subadditive Euclidean Functionals

By abstracting from the traveling salesman tour just a few of its basic properties, it is possible to suggest a very general result that provides information comparable to that given by the BHH theorem for a large number of problems of combinatorial optimization in Euclidean space.

Let L be a function that associates a real number to each finite subset $\{x_1, x_2, \ldots, x_n\} \subset R^d$. To spell out the most innocent properties of L that mimic the behavior of the TSP, we first note that for the TSP, L exhibits **homogeneity** and **translation invariance**; *i.e.*,

$$L(ax_1,\ldots,ax_n) = aL(x_1, x_2,\ldots,x_n) \text{ for all } \alpha > 0,$$

and

$$L(x_1 + y, x_2 + y,\ldots, x_n + y) = L(x_1, x_2,\ldots,x_n) \text{ for all } y \in R^d$$

The TSP's length also has some strong smoothness and regularity properties, but these turn out not to be too important for most purposes. All that is needed is that L be Borel measurable when viewed as a function from $\mathbf{R}^{nd}$ to $\mathbf{R}$. This condition is almost always trivial to obtain, but still one has to have it to be able to talk honestly about probabilities involving L.

Functions on the finite subsets of $\mathbf{R}^d$ that are measurable in the sense just described and that are homogeneous of order one and translation invariant are called **Euclidean functional**. These three properties are commonplace but bland. One should not expect to be able to prove much in such a limited context, but, with the addition of just a couple of other structural features, a rich and useful theory emerges.

The first additional property of the TSP functional that we consider is that it is **monotone** in the sense that

$$L(x_1, x_2,\ldots,x_n) \le L(x_1, x_2,\ldots,x_n, x_{n+1}) \ge 1, \text{ and } L(\phi) = 0$$

The final feature of the TSP functional that we abstract is the most substantial one. It expresses both the geometry of the underlying space and the fundamental suboptimality of one of the most natural TSP heuristics, the partitioning heuristic. If $\{Q_i\}_{i=1}^{m^d}$ is a partition of $[0, t]^d$ into smaller cubes of edge length t/m, the **sub-additive property** says there exists a constant B independent of *m* and *t* such that

$$L(\{x_1, x_2,\ldots,x_n\} \cap [0,t]^d) \le \sum_{i=1}^{m^d} L(\{x_1, x_2,\ldots,x_n\} \cap Q_i) + Btm^{d-1}$$

for all integers $m \ge 1$ and real $t \ge 0$.

Euclidean functional that satisfy equations will be called monotone sub-additive Euclidean functional. This class of processes seems to abstract the most essential features of the TSP that are needed for an effective asymptotic analysis of the functional applied to finite samples of independent random variables with values in $\mathbf{R}^d$. To see how sub-additive Euclidean functional arise, it is useful to consider two additional examples.

The first is the Steiner minimum tree. For any finite set of n points, $S = \{x_1, x_2,...,x_n\} \subset R^d$, a Steiner minimum tree for S is a tree T whose vertex set contains S such that the sum of the lengths of the edges in T is minimal over all such trees. Note that the vertex set of T may contain points not in S; these are called Steiner points. If $L_{ST}(x_1, x_2, ..., x_n)$ is the length of a Steiner tree of $x_1, x_2, ..., x_n$ and if we let $|e|$ be the length of an edge e, another way of defining L_{ST} is just

$$L_{ST}(S) = \min_T \left\{ \sum_{e \in T} |e| : T \text{ is a tree containg } S \subset R^d, S \text{ finite} \right\}.$$

The second example is closely related, yet it points out that the innocuous monotonicity property of the TSP can fail even in natural problems. The example we have in mind is the **minimum spanning tree**. For

$\{x_1, x_2, ..., x_n\} \subset R^d$, let $L_{MST}(x_1, x_2, ..., x_n) = \min \sum_{e \in T} |e|$, where the minimum is over all connected graphs T with vertex set $\{x_1, x_2, ..., x_n\}$. It is an easy matter to check that any minimizing graph must indeed be a spanning tree.

The functional L_{MST} is easily seen to be homogeneous, translation-invariant, and measurable. One can also check without much trouble that it is subadditive in the sense required above. Still, many simple examples such as the sets $S = \{(0,0), (0,2), (2,0), (2,2)\}$ and $S \cup \{(1, 1)\}$ will show that L_{MST} fails to be monotone. One should suspect that this failure is of an exceptional sort that would not have great influence on asymptotic behavior, and this suspicion is well justified. Still, the example puts us on warning that non-monotone functionals can require delicate considerations that axe not needed in cases that mimic the TSP more closely.

The main theorem of this section shows that the properties (1.1) through (1.4), together with a modest moment condition, are sufficient to determine the asymptotic behavior of $L(X_1, X_2,..., X_n)$ when the X_i are independent and identically distributed:

Theorem 1.2 Let L be a monotone subadditive Euclidean functional. If $\{X_i\}$, i=1, 2, ... , are independent random variables with the uniform distribution on $[0, 1]^d$, and Var $L(x_1 x_2, ..., x_n) < \infty$ for each $n \geq 1$, then as $n \to \infty$

$$L(X_1 X_2, ... X_n) / n^{(d-1)/d} \to \beta_{L,d}$$

with probability one, where $\beta_{L,d} \geq 0$ is a constant depending only on L and d.

The restrictions that this theorem imposes on a Euclidean functional are as few as one can reasonably expect to yield a generally useful limit theorem, and because of this generality the restriction to uniformly distributed random variables is palatable. Moreover, since many of the probabilistic models studied in operations research and computer science also focus on the uniformly distributed case, the theorem has immediate applications. Still, one cannot be long content with a theory confined to uniformly distributed random variables. Fortunately, with the addition of just a couple of additional constraints, the limit theory of subadditive Euclidean functionals can be extended to quite generally distributed variables.

To get to the essence of the extension, we first consider the case of random variables with a singular component, *i.e.*, variables such that there is a measurable set $E \subset R^d$ with Lebesgue measure zero such that $P(X_i \in E) > 0$. Random variables with a singular component provide a kind of geometrical worst case and hence provide a useful test case.

To deal with such random variables, we need to draw out three additional properties shared by many subadditive Euclidean functionals. The first of these, called **scale-boundedness**, holds provided there is a constant C such that

$$L(x_1, x_2, \ldots, x_n) \leq Ctn^{(d-1)/d}$$

for all $\{x_1, x_2, \ldots, x_n\} \subset [0,t]^d$. The second property, called **simple subadditivity**, holds provided there is a constant D such that

$$L(A_1 \cup A_2) \leq L(A_1) + L(A_2) + Dt$$

for any finite subsets A_1 and A_2 contained in $[0, t]^d$.

These two properties hold for many natural examples, and in particular they are easily checked for the TSP. The key benefit of scale-boundedness and simple subadditivity comes from their showing that the singular component of a distribution makes a contribution that is of lower order than that of the absolutely continuous component.

Lemma 1 Let L be a monotone subadditive Euclidean functional that is scale-bounded (8.5) and simply subadditive (8.6). If $\{X_i\}$ are independent identically distributed (*i.i.d.*) random variables and E is any bounded set of Lebesgue measure zero, then as $n \to \infty$

$$L(\{X_1, X_2, \ldots, X_n\}) \cap E / n^{(d-1)/d} \to 0$$

with probability one.

The last property we will require for the extension of this theorem to general distributions calls on a type of restricted converse of simple sub-additivity. *A* Euclidean functional *L* is called **upper-linear** provided we have the following: For every finite collection of cubes Q_j, where $1 \leq j \leq M$ and the edges of the Q_j are parallel to the axes, and for every infinite sequence x_i, where $1 \leq i < \infty$ and $x_i \in R^d$, we have

$$\sum_{j=1}^{m} L(\{x_1, x_2, \ldots, x_n\}) \cap Q_j \leq L(\{x_1, x_2, \ldots, x_n\}) \cap \cup_{j=1}^{M} Q_j + (n^{(d-1)/d})$$

1.4 Tail Probabilities

The theory just outlined has a number of extensions and refinements. The first of these that we consider is the work of Rhee and Talagrand on the behavior of the tail probabilities of the TSP and related functionals under the model of independent uniformly distributed random variables in the unit d-cube. It was observed that in dimension 2 Var (L_n) is bounded independently of n. This somewhat surprising result motivated the search for a more detailed understanding of the tail probabilities $P(L_n \geq t)$, and, in particular, it opened the question of determining if these probabilities might decay at the Gaussian rate $\exp(-ct^2/2)$.

After introducing several methods from martingale theory and interpolation theory that led to interesting intermediate results, Rhee and Talagrand (1989) finally provided a remarkable proof that in $d = 2$ the TSP (and many related functionals) do indeed have Gaussian tail bounds.

Theorem 1.3 Suppose f is a Borel measurable function that assigns to each finite subset $f \subset [0,1]^2$ a real value $f(F)$ such that

$$f(F) \leq f(F \cup x) \leq f(F) + \min(d(x,y) : y \in F).$$

There is then a constant $K = K_f$ such that if X_i are independent and uniformly distributed in $[0, 1]^2$, then the random variables defined by $U_n = f(\{X_1, X_2, \ldots, X_n\})$ satisfy $P(|U_n - E(U_n)| > t) \leq \exp(-t^2/K)$

The proof of this result rests on the systematic exploitation of the martingale versions of the Hoeffding and Prokhorov large-deviation inequalities together with a powerful bare-hands construction that helps articulate technical features of a random sample being well spread out.

1.5 TSP in Fractal Spaces

The BHH theorem may seem like a result that is wedded to $\mathbf{R}^d$; certainly the growth rate $n(d-1)/d$ emerges as a natural characteristic of the dimension. It gives a set of results that cast new light on the role of dimension in problems like the TSP by developing limit results for TSP tours in fractal spaces. Lalley considered finite random samples chosen uniformly (in an appropriate sense) from self-similar subsets of $\mathbf{R}^2$ that have Hausdorff dimension strictly less than two. Lalley obtained an almost-sure limit law for the length L_n of the shortest tour through $\{X_1, X_2, \ldots, X_n\}$, but now the normalizing denominator is no longer $\sqrt{n}$ but a power of n that depends on the structure of the self-similar set.

To give the flavor of the results obtained by Lalley, we will consider one specific example. A compact set K is said to be a strongly self-similar subset of $\mathbf{R}^2$ provided that there is a finite collection of transformations $T_i : K \to K$ so that

$$K = \bigcup_{i=1}^{m} T_i(K),$$

where each h T_i is a transformation that can be written as a strict affine contraction (with contraction factor $r_i < 1$) followed by a rigid motion. The strong self-similarity of K requires moreover that the intersections $T_i(K) \cap T_j(K)$, where $i \neq j$, be small in a technical sense that we will not detail.

Theorem 1.4 If X_i, $1 \leq i < \infty$, are independent random variables with the "uniform distribution" on a strongly self-similar set, if that has representation

$$K = \bigcup_{i=1}^{m} T_i(K),$$

where $\| T_i(x) - T_i(y) \| = r_i \| x - y \|$, then the length L_n of the shortest tour through the point set $\{x_i : 1 \leq i \leq n\}$ satisfies

$$\lim_{n \to \infty} L_n / n^{\theta} = C_K$$

with probability one, where $C_K > 0$ and $0 < \theta \leq 1/2$ is a constant uniquely determined by the similarity ratios r_i.

1.6 Minimal Spanning Trees

If $x = \{x_1, \ldots, x_n\}$ is a finite set of points in $\mathbf{R}^d$ for $d \geq 2$, , a minimal spanning tree (MST) t of x is a connected graph with vertex set x such that the sum of the edge lengths of t is minimal; *i.e.*,

$$\sum_{e \in t} |e| \min_{G} \sum_{e \in G} |e|,$$

where $|e| = |x_i - x_j|$ is the Euclidean length of the edge $e = (x_i, x_j)$ and the minimum is over all connected graphs G.

Minimal spanning trees are among the most studied objects in combinatorial optimization, and even the probability theory of MST is rather well developed. For example, in Steele (1989) one finds results that cover the basic almost-sure asymptotic behavior of the MST that parallels the BHH theorem and even deals with edge weights that are more complicated than those given by the basic edge lengths. In particular, it is shown that if X_i are *i.i.d.* with compactly supported density f and $0 < \alpha < d$, then

$$n^{-(d-\alpha)/d}\sum |e|^{\alpha} \to c(\alpha, d)\int_{R^d} f(x)^{(d-\alpha)/d}\,dx \quad \text{a.s.},$$

where c(α, d) depends only on α and *d*.

One peculiar aspect of this result is that it fails to cover the extreme case $\alpha = d$. This failure was particularly intriguing since the case $\alpha = d$ for $d = 2$ had been the source of an interesting conjecture by R. Bland, who by direct computational experience had been led to believe that if one takes the sum of the squares of the edges of the MST of a sample of points chosen randomly from the unit square, then as $n \to \infty$ the sum converges to a constant. This result is hinted at by the limit cited above, but the specifically desired result fell just outside of its domain.

An approach to such limit problems was taken that differs completely from the subadditivity-based analyses such as that mentioned in 1.2 or the more ad hoc method used in Steele (1989). The new approach is based on the study of an analog of the MST for an infiniteset of points, in particular the points of a Poisson process on all $\mathbf{R}^d$. The basic philosophy that guided the analysis is that because the empirical distributions of any independent uniform sequences look locally like a realization of the Poisson process, one should be able to relate the MST of a uniform sample to the MST of the (unbounded) Poisson process. Though this sounds a little like the "Poissonization" trick that goes back even as fax as Beardwood, Halton, and Hammersley, it is really radically different in concept and detail. To give a forward hint of that difference, note that we have no idea how to apply the method to the TSP. To sketch the basic idea, let $x = (x_i)$ be a finite or countably infinite subset of $R^d, d \geq 2$. We call x nice if (1) x is locally finite, *i.e.*, has only finitely many elements in bounded subsets of $\mathbf{R}^d$, and (2) the interpoint distances ($|x_j - x_i|$, $i < j$) are all distinct.

Now, given a pair (x, x) with x nice and $x \in x$, , we can define trees $t_m(x, x)$ with vertices from x as follows: Let $\xi_1 = x$, and let t_1 be the single vertex ξ_1. Let t_2 be the tree consisting of the vertex ξ_1 and the vertex $\xi_2 \in x \setminus \{\xi_1\}$ that is closest to ξ_1 in Euclidean distance, together with the edge (straight line segment) connecting ξ_1 and ξ_2. We then proceed inductively in a greedy fashion and define $t_m = t_m(_x, x)$ to be t_{m-1} together with a new edge $(\xi_{m'}, \xi_m)$, where $j_m \leq m-1$ and $\xi_m \in x \setminus \{\xi_1, \ldots, \xi_{m-1}\}$ are chosen so that the edge length $|\xi_m - \xi_{jm}|$ is minimal (over all possible edges connecting $t_{m-1}$to $x \setminus t_{m-1}$.) In the finite case where $n = |x| < \infty$, this procedure terminates with the tree $t_n(x, x)$, and in that circumstance the tree obtained does not depend on the choice of the starting vertex x.

When x is infinite, the situation is more complex. We write $t_\infty(x, x)$ for the set $\bigcup_n t_n(x, x)$, but some work is required to obtain useful structural information about this graph. In fact, it turns out to be technically useful to look at a different but related graph, described in the next lemma.

Lemma 2 Let $g = g(x)$ be the graph on an infinite nice vertex set x defined by taking (x_1, x_2) as an edge in g if it is an edge in either $t_\infty(x_1, x)$ or $t_\infty(x_2, x)$. Then the graph g is a **forest** and each component of g is an infinite tree.

The main object of Aldous and Steele (1992) is the random tree $\mathcal{T}$ constructed by taking a Poisson point process $\mathcal{N} = \{\eta_i\}$ of rate 1 in $\mathbf{R}^d$, letting $\mathcal{N}^0 = \mathcal{N} \cup \{0\}$.

Let $\mathcal{G} = g(\mathcal{N}^0)$ be the forest constructed as in the lemma, and finally take the connected graph $\overline{\mathcal{T}}$ to be the largest tree containing 0.

It is natural to conjecture that $\mathcal{G}$ is itself a tree with probability 1, but this possibility seems to be related to deep issues in continuum percolation, and the introduction of $\overline{\mathcal{T}}$ permits one to finesse that subtle issue.

Lemma 3 Let D be the degree of vertex 0 in $\overline{\mathcal{T}}$ and let $L_1, \dots, L_D$ be the lengths of the edges of $\overline{\mathcal{T}}$ incident to 0. We then have

1. $D \le b_d$, a constant,
2. $\mathbf{E}D = 2$, and
3. $l_d \equiv \frac{1}{2} \sum_i EL_i^d < \infty$

Now let $\mathcal{N}_n$ denote the point process consisting of n points $(\eta_i : 1 \le i \le n)$ that are independent and uniformly distributed on the unit cube $[0, 1]^d$. With probability 1, $\mathcal{N}_n$ is a nice subset of $\mathbf{R}^d$. Let $S_n = t_n(\eta_1, \mathcal{N}_n)$ be the minimal spanning tree of these n vertices. It is intuitive that $S_n|$, after a suitable rescaling, should converge locally to $\mathcal{G}$.

Theorem 1.5 (a) If $\{|e_i|\}$ denotes the set of lengths of the edges of $S_n|$, then

$$\sum |e_i|^d \xrightarrow{L^2} l_d \text{ as } n \to \infty.$$

(b) If $\Delta_{n,i}$ denotes the proportion of vertices of S_n with degree i, then for each i,

$$E\Delta_{n,i} \to P(D = i) \text{ as } n \to \infty.$$

The existence of limits in (b) was proved by different methods, where one also finds a proof of almost sure convergence. The analogous question in the model in which the cost assigned to each edge is independent and identically distributed is solved using "limit process" arguments such as those reviewed above.

1.7 Matching Problems

The problem of determining the least weight matching in a graph (*G, E*) for which there is a function that assigns a real value to the edges of *G* is a central problem of combinatorial optimization. The matching problem is intermediate in complexity between the MST problem and the TSP. The minimal matching problem cannot be solved by a naive greedy algorithm such as that used for the MST, but there is still a polynomial-time algorithm that does solve the problem—in contrast to the TSP.

There are two natural Euclidean versions of the matching problem. The simplest version considers $2n$ points in $\mathbf{R}^d$ and asks for the perfect matching of the set for which the total edge length is minimum. This matching problem has a theory that is almost perfectly parallel to the theory of the MST.

A more subtle set of issues arise when one instead considers two distinguished subsets $\{x_1, x_2, \ldots, x_n\}$ and $\{y_1, y_2, \ldots, y_n\}$. Here the probability theory becomes much trickier, and new phenomena arise. A first taste of the new behavior is given by the results, one of which we note below:

Theorem 1.6 If X_i and Y_i are independent and uniformly distributed in $\mathbf{R}^2$ for $1 \le i \le n$, then there are constants K_1 and K_2 such that

$$K_1\sqrt{n\log n} \le M_n = \min_{\sigma}\sum\left|X_i - Y_{\sigma(i)}\right| \le K_2\sqrt{n}\log n$$

with probability that approaches one as $n \to \infty$.

The $\sqrt{\log n}$ term in this result adds a new spin to the theory of Euclidean functionals for several reasons. First, it shows that the scaling arguments and self-similarity ideas that were useful for the TSP and the MST are no longer accurate enough to lead to the right orders. Still, the basic motivation remains substantially intact, and for $d \ge 3$ one again finds the characteristic growth rates of *n(d-1)/d*.

As an added impetus to the theory of Euclidean matching, there are some powerful connections with other parts of combinatorial optimization. It was through their work on bin packing and scheduling that Leighton and Shor were led to the investigation of the maximum length that must exist in a two-sample matching. The following example from their work again exhibits a curious logarithmic term:

Theorem 1.7 If X_i and Y_i are independent and uniformly distributed in $[0, 1]^2$ for $1 \le i \le n$, then there is a constant K such that

$$\min_{\sigma}\max_{1\le i\le n}\left|X_i - Y_{\sigma(i)}\right| \le K\sqrt{n}\,(\log n)^{3/4}$$

with probability that approaches one as $n \to \infty$.

In one of the most recent and most sweeping studies reviewed for this survey, has unified the two preceding results as well as many others. The details of that development are too substantial to review here, but it is possible to indicate some of the sources of the new power.

Begin by noting that there is a basic relation between the behavior of the two sample matching problems and the theory of empirical discrepancy. We recall that if $\overline{C}$ is any class of functions from [0, 1] to [0, 1], the **empirical discrepancy** associated with C is defined by

$$D_n(C) = \sup_{f \in C} \left| \sum \left(f(X_i) - \int_{R^d} f(x)dx \right) \right|$$

where the X_i are *i.i.d.* random variables with density *f*. Even before the work, it was well understood that there was a close connection between M_n and D_n when C is taken to be the class of Lipschitz functions. It was also known that the theory associated with D_n was closely linked to the epsilon entropy of the class *C*, and finally people were beginning to understand that the issues addressed by epsilon entropy could in some cases be more effectively addressed by an emerging method of majorizing measures.

Unfortunately, even defining all these terms would take us rather far afield, but hopefully it is meaningful to say that epsilon entropy is a tool for measuring the complexity of a class of functions by counting the number of epsilon balls that are needed to cover certain compact subsets, whereas the method of majorizing measures refines the theory of epsilon entropy to deal more effectively with aspects of inhomogeneity in *C*. The substantial achievement involves development of concrete tools of the theory of majorizing measures that make it directly useful in the theory of M_n and D_n and then to show how those tools could be used to deepen and unify the investigations.

1.8 Values of Constants

There is a long history of effort devoted to the determination of the limiting constraints in results like the BHH theorem. There is also a substantial literature that addresses the constraints associated with worst-case upper bounds.

The work provides recent progress on the worst-case bounds and provide surveys of the older literature which accounts for two particularly noteworthy contributions for TSP. The latter broke the $\sqrt{2}$ barrier in dimension 2 by showing that $\alpha TSP, 2 \leq 0.984\sqrt{2}$, whereas obtained the best bounds in general dimension by means of a powerful technique that uses an infinite number of translations of quantizers other than cubical cylinders.

After the constraints on the TSP, the best-studied value is $\alpha_{RST,d}$, the constant associated with the worst-case length of a rectilinear Steiner minimum tree in the unit *d*-cube. As proved that αRST, 2 = 1, which is significant in that it is the only nontrivial worst-case constant for which we have an exact expression. The problem of determining $\alpha_{\mathrm{RST,\,d}}$ in higher dimensions is still open, the bounds being

$$\max\{1, d/(4e)\} \leq \alpha RST, d \leq d4^{(1-d)/d} \text{ for } d \geq 1$$

Other than these bounds, little progress concerning the $\beta_{L,d}$ was made which developed exact expressions for the probabilistic minimum spanning tree and matching constants as d gets large. Specifically, they showed that $\beta_{MST,d} \sim \sqrt{d/2\pi e}$ and $\beta_{M,d} \sim \left(1/2\sqrt{d/2\pi e}\right)$ as $d \to \infty$.

The crowning achievement concerning the probabilistic constants was the determination of an exact expression for $\beta_{MST,d}$ for all $d \geq 2$. The expression for $\beta_{MST,d}$ is a series expansion, each term of which requires a rather difficult integration. Still, the first few terms of the series in dimension 2 have been computed to yield a numerical lower bound of $\beta_{MST,2} \geq 0.599$, which agrees well with experimental data. We note that the proof employed by Avram and Bertsimas relies strongly on the fact that a greedy construction is guaranteed to yield an MST. Unfortunately, such a construction fails for many objects of interest, including the TSP. It is therefore evident that entirely new methods probably will be required to develop exact expressions for constants such as $\beta_{TSP,d}$ that arise from NP-hard problems.

As a final note, we return to the rectilinear Steiner minimum tree problem. Bern showed that, for a Poisson process with intensity N on the unit square, the value of the rectilinear MST constant is at least that of the rectilinear Steiner tree constant plus 0.0014. This separation of the constants was increased where extended results include points uniformly distributed in the unit square.

1.9 Central Limit Problem

The recent work provides enticing progress on the important and long-standing problem of providing a central limit theory (CLT) for Euclidean functionals. Their method falls short of providing a CLT for the MST problem or the TSP, but it provides very useful CLTs for somewhat easier problems, including the kth nearest neighbor problem, the Delaunay triangulation, and the length of the Voronoi diagram.

There axe two keys to the approach used by Avram and Bertsimas. The first is the observation of the applicability of the relatively recent CLT for dependent random variables which deals with the dependence relations within a collection of random variables through the structure of a dependency graph G_n and offers an explicit Berry-Essen type bound where the maximal degree D_nG_n plays a critical role. The second key observation is that one achieves a considerable simplification of the CLT problem by first conditioning on the event A_n that in the subdivision of [0, 1] into subcubes of approximate size $n/\log n$ each subcube contains at least one and at most $e \log n$ of the sample points. The specifics of the problems handled by the Avram-Bertsimas method are such that once one conditions on A_n, one finds a problem that is within range of the Baldi-Rinott CLT. Any given problem has a number of details that must be directly resolved.

1.10 Worst-Case Growth Rates

One engaging aspect of the asymptotic theory of combinatorial optimization of point functionals is the persistent similarity between the probabilistic rates of growth that have been reviewed and the behavior of the corresponding functionals in worst-case settings.

As a primary example of a worst-case growth rate, consider the **worst-case length** of an optimal traveling salesman tour in the unit d-cube:

$$PTSP(n) = \max_{\substack{S \subset [0,1]^d \\ |S| = n}} \min_{T} \left(\sum_{e \in T} l|e| : T \text{ is a tour of } S \right)$$

In words, $\varrho_{TSP}(n)$ is the maximum length over all n point sets in $[0, 1]^d$ that an optimal traveling salesman tour can attain. There is absolutely no probability theory here, for the point sets and tours are deterministic, yet there is a theory for this function that parallels the BHH theorem as closely as one could hope.

Theorem 1.8 As $n \rightarrow 8$,

$$\rho TSP(n) \sim \alpha TSP, d\ n^{(d-1)/d},$$

where $\alpha_{TSP,d} > 0$ is a constant depending only on the dimension d.

The proof of this theorem and several related results all call on the subadditivity and smoothness properties of the underlying functional. The following elementary lemma often helps focus one's investigation of these problems by putting a finger on one clear set of sufficient conditions.

Lemma 4 If $\varrho(1) = 0$ and there exists a constant $c \geq 0$ such that for all integers $m \geq 1$ and $k \geq 1$, we have

1. $\rho(n+1) \geq \rho(n) + cn^{-1/d}$ *and*
2. $m^{d-1}\rho(k) - m^{d-1}k^{(d-1)/d}r(k) \leq \rho(m^d k)$,

where $r(k) \rightarrow 0$ as $k \rightarrow \infty$, then as $n \rightarrow \infty$, $\rho(n) \sim \alpha n^{(d-1)/d}$ for a positive constant α.

For the analysis of any particular problem, there is almost always serious work to be done to make the problem amenable to this lemma or its variant. Still, the lemma has already proved its effectiveness in a number of problems, including minimum matchings, minimum spanning trees, greedy matchings and rectilinear Steiner minimum trees.

1.11 Combinatorial Optimization Problems

1.11.1 Unified Model

The model with this appealing property is the unconstrained quadratic binary programming problem, accompanied by the device of introducing quadratic infeasibility penalty functions to handle constraints. Not only is this model capable of representing many "special case" problem classes, but it can be advantageously exploited by adaptive memory (tabu search) metaheuristics and associated evolutionary (scatter search) methods. Computational outcomes disclose the effectiveness of this combined modeling and solution approach for problems from a diverse collection of challenging settings. The unconstrained quadratic program can be written in the form:

$$\text{UQP: } \min f(x) = xQx$$

where Q is an n by n matrix of constants and x is an n-vector of binary variables. UQP is notable for its ability to represent a significant variety of important problems. The application potential of UQP is substantially greater than this, however, due to reformulation methods that enable certain constrained models to be re-cast in the form of UQP. Nonetheless, few applications of this idea appear in the literature. Our purpose here, based on extensive experience with a wide variety of problems, is to establish that such reformulation into the UQP format is not merely a representational novelty, but a unified framework of practical consequences.

1.11.2 Transformation to xQx

Many practical combinatorial optimization problems can be modeled as constrained optimization problems of the form

$$\min x_0 = xQx$$

subject to

$$Ax = b, \quad x \text{ binary}$$

The foregoing model accommodates both quadratic and linear objective functions since the linear case results when Q is a diagonal matrix (observing that $x_j^2 = x_j$ when x_j is a 0-1 variable). Problems with inequality constraints can also be put into this form by introducing so-called *slack variables* to convert the inequalities into equations, and representing these bounded slack variables by a binary expansion. These constrained quadratic optimization models are then converted into equivalent UQP models by adding a quadratic infeasibility penalty function to the objective function as an alternative to explicitly imposing the constraints $Ax = b$. The general approach to such re-casting, which we call transformation # 1, is given below:

Transformation 1. We choose a positive scalar P, to yield

$$x_0 = xQx + P(Ax-b)^t(Ax-b)$$
$$= xQx + xDx + c$$
$$= x\hat{Q}x + c$$

where the matrix *D* and the additive constant *c* result directly from the matrix multiplication indicated. We can drop the additive constant, whereupon the equivalent unconstrained version of our constrained problem become

$$UQP(PEN): \min x\hat{Q}x,\ x\ binary$$

From a theoretical standpoint, a suitable choice of the penalty scalar P can always be chosen so that the optimal solution to UQP(PEN) is the optimal solution to the original constrained problem. From a practical standpoint, however, experience has shown that penalty-based conversions in other settings have uniformly proved to be highly unstable, engendering numerical difficulties and poor solution performance when the penalties are large, and producing invalid representations of the original problem when the penalties are smaller. Finding a proper trade-off between penalty size (and the design of a method to exploit the penalized representation) has turned out to be feasible only in the case of linear and convex programming domains, where penalty considerations are much simpler than in combinatorial optimization. By contrast, however, our experience with penalty-based representations of the UQP model for combinatorial optimization problems has shown them to be easy to work with and highly robust. As reported, valid and computationally stable penalty values can be found without difficulty and a wide range of such values work well.

In addition to the modeling possibilities introduced by Transformation 1, a very important special class of constraints that arise in many applications can be handled by an alternative approach, given below.

Transformation 2. This approach is convenient for problems with considerations that isolate two specific alternatives and prohibit both from being chosen. That is, for a given pair of alternatives, one or the other but not both *may* be chosen. If x_j and x_k are binary variables denoting whether or not alternatives *j* and *k* are chosen, the standard constraint that allows one choice but precludes both is:

$$x_j + x_k \le 1$$

Then, for a positive scalar *P*, adding the penalty function Px_jx_k to the objective function is a simple alternative to imposing the constraint in a traditional manner. For problems with a linear objective function, the scalar P (with respect to transformation # 2) can be chosen as small as the largest objective function coefficient. This penalty function has sometimes been used by to convert certain optimization problems on graphs into an equivalent UQP model as referenced in the previous section. Its potential application, however, goes far beyond these settings as demonstrated in this paper. Variable upper bound constraints of the form $x_{ij} \le y_i$ can be accommodated by Transformation 2 by first replacing each y_i variable by $1 - y_i'$, where y_i' is the complementary variable that equals 1 when $y_i = 0$ and equals 0 when $y_i = 1$. The opportunity to employ this modeling device in the context of Transformation 2 makes it possible to conveniently model a variety of additional problem types.

The constraint associated with Transformation # 2 appears in many important applications which leads us to single it our here as an important alternative to Transformation # 1. We note, however, that many other problem specific special cases exist that can be employed to quickly yield quadratic equivalent representations.

1.11.3 Examples

Before highlighting some of the problem classes to which we have successfully applied the foregoing transformation approaches, we give two small examples from classical NP-hard problem settings to provide concrete illustrations.

Example 1: Set Partitioning. The classical set partitioning problem is found in applications that range from vehicle routing to crew scheduling. As an illustration, consider the following small example:

$$\min x_0 = 3x_1 + 2x_2 + x_3 + x_4 + 3x_5 + 2x_6$$

subject to

$$x_1 + x_3 + x_6 = 1$$

$$x_2 + x_3 + x_5 + x_6 = 1$$

$$x_3 + x_4 + x_5 = 1$$

$$x_1 + x_2 + x_4 + x_6 = 1$$

and x binary. Applying Transformation 1 with P =10 gives the equivalent UQP model:

$$UQP(PEN): \min x\hat{Q}x,\ x \text{ binary}$$

where the additive constant, c, is 40 and

$$\hat{Q} = \begin{bmatrix} -17 & 10 & 10 & 10 & 0 & 20 \\ 10 & -18 & 10 & 10 & 10 & 20 \\ 10 & 10 & -29 & 10 & 20 & 20 \\ 10 & 10 & 10 & -19 & 10 & 10 \\ 0 & 10 & 20 & 10 & -17 & 10 \\ 20 & 20 & 20 & 10 & 10 & -28 \end{bmatrix}$$

Solving UQP(PEN) by the Tabu Search method of Glover *et al.* [17, 18] we obtain an optimal solution $x_1 = x_5 = 1$, (all other variables equal to 0) for which $x_0 = 6$. In the straightforward application of Transformation 1 to this example, it is to be noted that the replacement of the original problem formulation by the UQP(PEN) model did not involve the introduction of new variables. In many applications, Transformation 1 and Transformation 2 can be used in concert to produce an equivalent UQP model, as demonstrated next.

Example 2: The K-Coloring Problem:

Vertex coloring problems seek to assign colors to nodes of a graph such that adjacent nodes are assigned different colors. The K-coloring problem attempts to find such a coloring using exactly K colors. A wide range of applications, ranging from frequency assignment problems to printed circuit board design problems, can be represented by the K-coloring model. Such problems can be modeled as satisfiability problems using the assignment variables as follows:

Let x_{ij} to be 1 if node *i* is assigned color *j*, and to be 0 otherwise. Since each node must be colored, we have

$$\sum_{j=1}^{K} x_{ij} = 1 \quad i = 1, \ldots, n \tag{1}$$

where n is the number of nodes in the graph. A feasible coloring requires that adjacent nodes are assigned different colors. This is accomplished by imposing the constraints

$$x_{ip} + x_{jp} \leq 1 \quad p = 1, \ldots, K \tag{2}$$

for all adjacent nodes (*i, j*) in the graph.

This problem can be re-cast into the form of UQP by using Transformation 1 on the assignment constraints of (1) and Transformation 2 on the adjacency constraints of (2). No new variables are required. Since the model of (1) and (2) has no explicit objective function, any positive value for the penalty, P, will do. The following example gives a concrete illustration of the re-formulation process.

Consider the following graph and assume we want find a feasible coloring of the nodes using 3 colors.

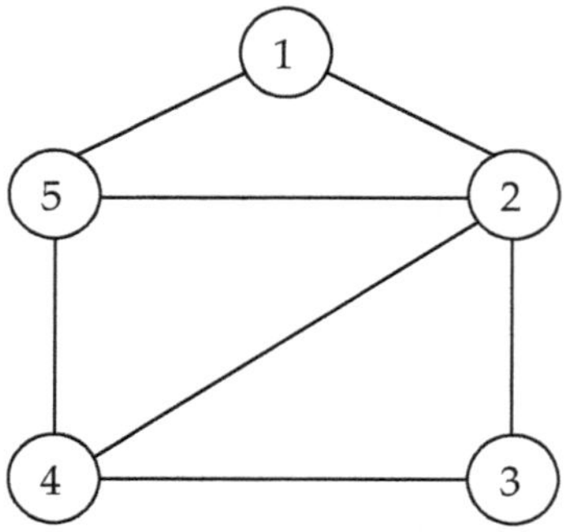

Our satisfiablity problem is that of finding a solution to:

$$x_{i1} + x_{i2} + x_{i3} = 1 \quad i = 1, 5 \tag{3}$$

$$x_{ip} + x_{jp} \leq 1 \quad p = 1, 3 \tag{4}$$

(for all adjacent nodes *i* and *j*)

In this traditional form, the model has 15 variables and 26 constraints. To recast this problem into the form of UQP, we use Transformation 1 on the equations of (3) and Transformation 2 on the inequalities of (4). Arbitrarily choosing the penalty P to be 4, we get the equivalent problem:

$$UQP(Pen): \min x\hat{Q}x$$

where the $\hat{Q}$ matrix is:

$$\hat{Q} = \begin{bmatrix} -4 & 4 & 4 & 4 & 0 & 0 & 0 & 0 & 0 & 0 & 0 & 0 & 4 & 0 & 0 \\ 4 & -4 & 4 & 0 & 4 & 0 & 0 & 0 & 0 & 0 & 0 & 0 & 0 & 4 & 0 \\ 4 & 4 & -4 & 0 & 0 & 4 & 0 & 0 & 0 & 0 & 0 & 0 & 0 & 0 & 4 \\ 4 & 0 & 0 & -4 & 4 & 4 & 4 & 0 & 0 & 4 & 0 & 0 & 4 & 0 & 0 \\ 0 & 4 & 0 & 4 & -4 & 4 & 0 & 4 & 0 & 0 & 4 & 0 & 0 & 4 & 0 \\ 0 & 0 & 4 & 4 & 4 & -4 & 0 & 0 & 4 & 0 & 0 & 4 & 0 & 0 & 4 \\ 0 & 0 & 0 & 4 & 0 & 0 & -4 & 4 & 4 & 4 & 0 & 0 & 0 & 0 & 0 \\ 0 & 0 & 0 & 0 & 4 & 0 & 4 & -4 & 4 & 0 & 4 & 0 & 0 & 0 & 0 \\ 0 & 0 & 0 & 0 & 0 & 4 & 4 & 4 & -4 & 0 & 0 & 4 & 0 & 0 & 0 \\ 0 & 0 & 0 & 4 & 0 & 0 & 4 & 0 & 0 & -4 & 4 & 4 & 4 & 0 & 0 \\ 0 & 0 & 0 & 0 & 4 & 0 & 0 & 4 & 0 & 4 & -4 & 4 & 0 & 4 & 0 \\ 0 & 0 & 0 & 0 & 0 & 4 & 0 & 0 & 4 & 4 & 4 & -4 & 0 & 0 & 4 \\ 4 & 0 & 0 & 4 & 0 & 0 & 0 & 0 & 0 & 4 & 0 & 0 & -4 & 4 & 4 \\ 0 & 4 & 0 & 0 & 4 & 0 & 0 & 0 & 0 & 0 & 4 & 0 & 4 & -4 & 4 \\ 0 & 0 & 4 & 0 & 0 & 4 & 0 & 0 & 0 & 0 & 0 & 4 & 4 & 4 & -4 \end{bmatrix}$$

Solving this unconstrained model, xQx, yields the feasible coloring:

$$x_{12}\ x_{22}\ x_{33}\ x_{41}\ x_{53} = 1 \quad \textit{all other} \quad x_{ij} = 0$$

This approach to coloring problems has proven to be very effective for a wide variety of coloring instances from the literature. Later in this paper we present some computational results for several standard *k*-coloring problems. An extensive presentation of the *xQx* approach to a variety of coloring problems, including a generalization of the K-coloring problem considered here, is given in Kochenberger, Glover, Alidaee and Rego.

1.11.4 Solution Approaches to UQP

Due to its computational challenge and application potential, UQP has been the focus of a considerable number of research studies in recent years, including both exact and heuristic solution approaches. Recent papers report on the branch and bound (exact) approaches as well as a variety of modern heuristic methods including simulated annealing, genetic algorithms, tabu search, and scatter search. (See [27] for references to these and other works.) Each of these approaches exhibits some degree of success. However, the exact methods degrade rapidly with problem size, and have meaningful application to general UQP problems with no more than 100 variables. For larger problems, heuristic methods are usually required. Several proposed heuristics, including the DDT method of Boros, Hammer and Sun and the "one-pass" procedures of Glover, Alidaee, Rego and Kochenberger have proven to be effective in certain instances. Two methods we have found to be particularly successful for a wide variety of problems are based on tabu search and on the related evolutionary strategy scatter search. In the following section we highlight our tabu search approach which was used to produce the computational results referenced later in this paper.

Although not pursued by us here, we note that an alternative approach is to solve UQP as a continuous non-linear optimization problem within the unit cube. This allows other heuristic/approximation methods based on continuous optimization methodologies to be applied.

1.11.4.1 Tabu Search

Our TS method for UQP is centered around the use of strategic oscillation, which constitutes one of the primary strategies of tabu search. The variant of strategic oscillation we employ may be sketched in overview as follows:

The method alternates between constructive phases that progressively set variables to 1 (whose steps we call "add moves") and destructive phases that progressively set variables to 0 (whose steps we call "drops moves"). To control the underlying search process, we use a memory structure that is updated at *critical events*, identified by conditions that generate a subclass of locally optimal solutions. Solutions corresponding to critical events are called *critical solutions*.

A parameter *span* is used to indicate the amplitude of oscillation about a critical event. We begin with *span* equal to 1 and gradually increase it to some limiting value. For each value of *span,* a series of alternating constructive and destructive phases is executed before progressing to the next value. At the limiting point, *span* is gradually decreased, allowing again for a series of alternating constructive and destructive phases. When *span* reaches a value of 1, a *complete span cycle* has been completed and the next cycle is launched.

Information stored at critical events is used to influence the search process by penalizing potentially attractive add moves (during a constructive phase) and inducing drop moves (during a destructive phase) associated with assignments of values to variables in recent critical solutions. Cumulative critical event information is used to introduce a subtle long-term bias into the search process by means of additional penalties and inducements similar to those discussed above. A complete description of the framework for the method is given in Glover, Kochenberger, Alidaee and Amini.

1.11.4.2 Computational Experience

Our results of applying the tabu search and associated scatter search metaheuristics to combinatorial problems recast in UQP form have been uniformly attractive in terms of both solution quality and computation times. Although our methods are designed for the completely general form of UQP, without any specialization to take advantage of particular types of problems reformulated in this general representation, our outcomes have typically proved competitive with or even superior to those of specialized methods designed for the specific problem structure at hand. Our broad base of experience with UQP as a modeling and solution framework includes a substantial range of problem classes including:

- Quadratic Assignment Problems
- Capital Budgeting Problems
- Multiple Knapsack Problems
- Task Allocation Problems (distributed computer systems)
- Maximum Diversity Problems
- P-Median Problems
- Asymmetric Assignment Problems
- Symmetric Assignment Problems
- Side Constrained Assignment Problems
- Quadratic Knapsack Problems
- Constraint Satisfaction Problems (CSPs)
- Discrete Tomography Problems
- Set Partitioning Problems
- Set Packing Problems

- Warehouse Location Problems
- Maximum Clique Problems
- Maximum Independent Set Problems
- Maximum Cut Problems
- Graph Coloring Problems
- Number Partitioning Problems
- Linear Ordering Problems
- Clique Partitioning Problems
- SAT problems

We are currently solving problems via UQP with more than 13,000 variables in the quadratic representation and are working on enhancements that will permit larger instances to be solved. For each of the problem classes listed above we have considerable computational experience showing that this approach consistently produces high quality solutions within very modest computational time. Below we present some representative results for two of the problem classes.

1.11.4.3 K-Coloring Results

Earlier we seen that to test the potential attractiveness of UQP modeling and solution approach to K-coloring problems, 15 standard test problems were recast into form of UQP and solved by tabu search method. Here in table, the description of problems and results are displayed. All computations were carried out on a 1.7 gigahertz PC.

Table 1.1: K-coloring Test Problems

# *ID*	# *Vertices*	# *Edges*	*xQx K*	*xQx Variables*	*xQx feasible*	*xQx Time*
Myciel3	11	20	4	44	Yes	< 1 sec
Myciel4	23	71	5	115	Yes	< 1 sec
Myciel5	47	236	6	282	Yes	< 1 sec
Myciel6	95	755	7	665	Yes	< 1 sec
Myciel7	191	2360	8	1528	Yes	< 1 sec
Anna	138	493	11	1518	Yes	47 sec
David	87	406	11	957	Yes	1 min, 13sec
Huck	74	301	11	814	Yes	2 sec
Jean	80	254	10	800	Yes	< 1 sec
Games120	120	638	9	1080	Yes	< 1 sec
Queen5_5	25	160	5	125	Yes	< 1 sec
Queen6_6	36	290	7	252	Yes	< 1 sec

# ID	# Vertices	# Edges	xQx K	xQx Variables	xQx feasible	xQx Time
Queen7_7	49	476	7	343	Yes	< 1 sec
Queen8_12	96	1368	12	1162	Yes	< 1 sec
Queen8_8	64	728	9	576	Yes	< 1 sec

The first four columns of Table 1.1 indicate the problem identifier along with the size of the graphs and the number of colors (K) to be used. The last three columns give the number of variables involved, whether or not a feasible coloring was found utilizing K colors, and the time our tabu search method took to find a solution. Note that feasible colorings (solutions) were quickly found in all 15 cases. In fact, the solutions shown in Table 1.1 are known to be optimal.

1.11.4.4 Max 2-SAT Results

Several authors have established the connection between SAT problems and nonlinear penalty functions. The special case of Max 2-SAT is particularly well suited for this approach as it leads naturally to an xQx representation. Our experience, as shown below, indicates that this is a very attractive way to approach this class of problems.

For a 2-SAT problem, a given clause could have zero, one, or two negations, each with a corresponding (classical) linear constraint. Each linear constraint, in turn, has an exact quadratic penalty that serves as an alternative to the linear constraint. The three possibilities and their constraint/penalty pairs are:

a) No negations:

   Classical constraint: $x_i + x_j \geq 1$

   Exact Penalty: $(1 - x_i - x_j + x_i x_j)$

b) One negation:

   Classical constraint: $x_i + \bar{x}_j \geq 1$

   Exact Penalty: $(x_j - x_i x_j)$

c) Two negations:

   Classical constraint: $\bar{x}_i + \bar{x}_j \geq 1$

   Exact Penalty: $(x_i x_j)$

It is easy to see that the quadratic penalties shown are zero for feasible solutions and positive for infeasible solutions. Thus, these special penalties can be used to readily construct a penalty function of the form of xQx (simply by adding the penalties together) which we seek to minimize. We have found this approach to be very effective on a variety of test problems. Table 1.2 shows the results we obtained via this approach on a set of test problems from the literature.

Table 1.2: Problems from Borchers & Furman.

n	*m*	*Best known solution*	*xQx solution*	*xQx time*	*Maxsat*[3] *solution*	*Maxsat time*
50	100	4	4	< 1	4	.4
50	150	8	8	< 1	8	1.5
50	200	16	16	< 1	16	116.2
50	250	22	22	< 1	22	652.4
50	300	32	32	< 1	32	8,763
50	350	41	41	< 1	NA	> 12 hr
50	400	45	45	< 1	NA	> 12 hr
50	450	63	63	< 1	NA	> 12 hr
50	500	66	66	< 1	NA	> 12 hr
100	200	5	5	< 2	5	3.2
100	300	15	15	< 2	15	13,770
100	400	29	29	< 2	NA	> 12 hr
100	500	44	44	< 2	NA	> 12 hr
100	600	?	65	< 2	NA	> 12 hr
150	300	4	4	< 3	4	4.1
150	450	22	22	< 3	NA	> 12 hr
150	600	38	38	< 3	NA	> 12 hr

As shown in the Table 1.2, by re-casting each Max 2-SAT instance into the form of xQx and solving the resulting unconstrained quadratic binary program with our Tabu Search heuristic, we were able to find best known solutions very quickly to all test problems considered. By way of contrast, the method of Borchers and Furman took a very long time on several problems and was unable to find best known results for several instances in the allotted 12 hour time limit. In addition to the problems of Table 1.2, we have successfully applied this approach to randomly generated problems with as many as 1000 variables and more than 10,000 clauses where best known results are found in roughly one minute of computation time.

The results shown in Tables 1.1 and 1.2 serve as strong evidence of the attractiveness of the xQx approach for the problems considered. Considering both solution quality and the time taken to produce these solutions, this approach is very competitive with special purpose methods constructed specifically for vertex coloring and max 2-Sat problems. We note in passing that similar performance relative to special purpose methods has been obtained for the other problem classes singled out earlier in the paper as well.

We have demonstrated how a variety of disparate combinatorial problems can be solved by first re-casting them into the common modeling framework of the unconstrained quadratic binary program. Once in this unified form, the problems can be solved effectively by adaptive memory tabu search metaheuristics and associated evolutionary (scatter search) procedures.

Our findings challenge the conventional wisdom that places high priority on preserving linearity and exploiting specific structure. Although the merits of such a priority are well-founded in many cases, the UQP domain appears to offer a partial exception. In forming UQP(PEN), we destroy any linearity that the original problem may have exhibited. Moreover, any exploitable structure that may have existed originally is "folded" into the $\hat{Q}$ matrix, and the general solution procedure we apply takes no advantage of it. Nonetheless, our solution outcomes have been remarkably successful, yielding results that rival the effectiveness of the best specialized methods.

This combined modeling/solution approach provides a unifying theme that can be applied in principle to all linearly constrained quadratic and linear programs in bounded integer variables, and the computational findings for a broad spectrum of problem classes raises the possibility that similarly successful results may be obtained for even wider ranges of problems. As our methods for UQP continue to improve with ongoing research, the UQP model offers a representational tool of particular promise.

1.12 Classification of Integer Programming Problems

In integer programming and the related field of combinatorics, it is common to differentiate between "easy" and "difficult" problems. To develop a method of classification, just four concepts are needed:

1. A class C of legitimate problems to which the theory applies
2. A nonempty subclass $C_A \subset C$ of "easy" problems
3. A nonempty subclass $C_B \subset C$ of "difficult" problems
4. A relation "not more difficult than" between pairs of legitimate problems

This immediately leads to:

Proposition (Reduction Lemma) Suppose that P and Q are two legitimate problems.

If Q is "easy" and P is "not more difficult than" Q, then P is "easy".

If P is "difficult" and P is "not more difficult than" Q, then Q is "difficult".

The class of legitimate problems is commonly defined as decision problems that have a YES-NO answer. Thus an *optimization problem* max{*cx* : *x S*}

for which an instance consists of {*c* and a 'standard' representation of *S*} is replaced by a *decision problem*:

Is there an $x \varepsilon S$ with value $cx \geq k$?

for which an instance of {*c*, a 'standard' representation of *S*, and an integer *k*}.

Definition 1: NP is the class of decision problems with the property that for any instance for which the answer is YES, there is a "short" (polynomial) procedure to validate the YES.

Definition 2: P is the class of decision problems in NP for which there exists a polynomial algorithm.

1.12.1 NP-Complete

Formally, NP-complete is a notion for so called recognition (or decision) problems, *i.e.*, problems defined by a question for which the only two possible answers are a YES or a NO. It is defined with respect to polynomial reductions. From Nemhauser and Wolsey (page 132) "X ε NP is said to be NP-complete if all problems in NP can be polynomially reduced to X." The first set of problems in this class is shown in Figure.

NP-hard problems are usually optimization problems whose recognition version is NP-complete. For example the TSP-optimization is NP-hard because its TSP-recognition version is NP-complete (TSP-rec is as follows: given *k* is there a tour of length ≤ *k* ?). Nemhauser and Wolsey say "A problem is NP-hard if there is an NP-complete problem that can be polynomially reduced to it." Thus if a problem is NP-hard it is at least as difficult as any NP-complete problem.

1.12.2 Reductions vs. Transformations

"P1 polynomially *reduces* to P2" if P1 can be solved in polynomial time using an algorithm for P2 as a subroutine (each call to the subroutine being counted as 1, so the number of such calls has to be polynomial). This is the Turing reduction concept. "P1 polynomially *transforms* to P2" if we can do the same as above but with the limit of only one call to the subroutine.

1.12.3 Polynomial vs. Exponential Growth

To understand the significance of polynomially bounded algorithms as a class, let us consider the remaining algorithms, those that violate all polynomial bounds. We usually refer to these as *exponential* algorithms, because 2*n* is the paradigm of nonpolynomial rates of growth. Other examples of exponential rates of growth are *kn* (any fixed *k* > 1), *n*!, 2*n*2, *nn* and *n*log *n*. It is obvious that, when the size of the input grows, any polynomial algorithm will eventually become more efficient than any exponential one as shown in Table 1.3.

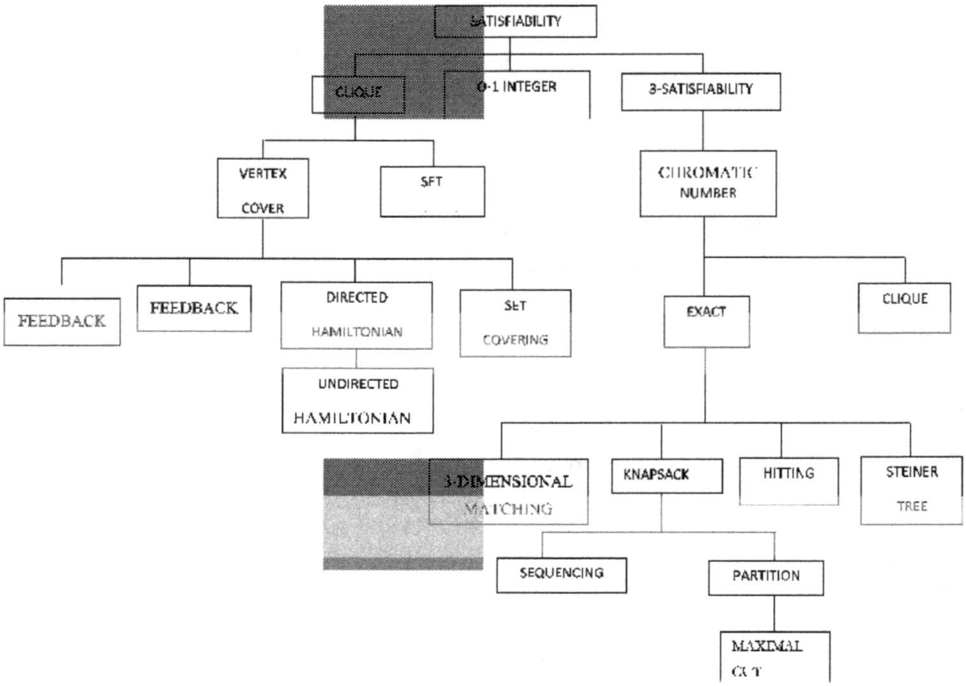

Table 1.3: Growth of Polynomial and Exponential Functions

Function	*Corresponding Values*		
n	10	100	1000
$n \log n$	33	664	9966
n^3	1000	1,000,000	10^9
$10^6 n^8$	10^{14}	10^{22}	10^{30}
2^n	1024	1.27×10^{30}	1.05×10^{301}
$n^{\log n}$	2099	1.93×10^{13}	7.89×10^{29}
$n!$	3,628,800	10^{158}	4×10^{2567}

Another positive feature of polynomial algorithms is that, in a sense, they take better advantage of technological advances. For example, each time a technological breakthrough increases the speed of computers tenfold, the size of the largest instance a polynomial algorithm can solve in a fixed amount of time will be multiplied by a constant between 1 and 10. In contrast, an exponential algorithm will experience only an *additive* increase in the size of the instance it can solve in a fixed amount of time (Table 1.4). Finally, it should be noted that polynomial algorithms have nice "closure" properties: (1) they may be combined to solve special cases of the same problem; (2) they may invoke another polynomial algorithm as a "subroutine" with the resulting algorithm still being polynomial.

Table 1.4: Polynomial-time Algorithms take better Advantage of Technology

Function	*Size of instance solved in one day*	*Size of instance solved in one day in a computer 10 times faster*
n	10^{12}	10^{13}
$n \log n$	0.948×10^{11}	0.87×10^{12}
n^2	10^6	3.16×10^6
n^3	10^4	2.15×10^4
$10^8 n^4$	10	18
2^n	40	43
10^n	12	13
$n^{\log n}$	79	95
$n!$	14	15

When differentiating between polynomial and exponential algorithms, special care must be taken when the time bound involves the numerical input of the problem. For example, given a directed graph $G = (V, A)$ with arc capacity $c(i,j)$ for arc (i,j) ε A, a well-known labeling algorithm for finding the maximum flow from some node s to some node t has time bound $0((\varepsilon\ (i,j)\ \varepsilon\ A\ c(i,j)) \bullet |A|)$.

The asymptotic growth of this function looks at first glance perfectly polynomial, because there is no obvious exponentiation or equivalent operation. Nevertheless, it is not a polynomial bounded function in terms of the size of the input. This follows because the complexity is expressed in terms of the numerical input of the instance. More generally, suppose that in a network all arcs have capacity $2^{|A|}$. We could encode this instance of the max-flow problem by listing $|A|$ integers, each $|A|$ bits long; thus the size of this instance is $0(|A|^2)$. However, the bound now becomes $0(2^{|A|} \bullet |A|^2)$, not a polynomial function of the size of the instance. The term *pseudopolynomial* is used for algorithms with time bounds such as this.

1.12.4 Polynomial vs. Exponential Algorithms

Let j be the largest integer in a given problem instance of, say, an IP. Because the binary encoding of j requires a string of length $O(\log \omega)$, an algorithm that requires ε steps is at least exponential.

The algorithm is exponential simply because the number of bits needed to encode j is exponential with respect to $\log(\omega)$; *i.e.*, the number of bits $y \sim \log(\omega)$, so $\omega \sim 2y$. A classic example is the 0-1 knapsack problem. Recall that the objective is to pack a set of items of highest total value into a knapsack of size b; that is,

$$\max\left\{\sum_{j=1}^{n} v_j x_j : \sum_{j=1}^{n} w_j x_j \le b,\ x_j \in \{0,1\} \text{ for all } j\right\}$$

A well-known dynamic programming algorithm to solve the problem has complexity $O(nb)$ where n is the number of different items to consider. Because it takes $\log(b)$ bits to encode the integer b, this algorithm is exponential *ow*, if n can be expressed as a polynomial function of b (*e.g.*, $n \sim b4$ so $O(nb) = O(n1.25)$, such an algorithm would be polynomial because in order to encode the weight (w) and the value (v) of each of the n items, you'll need at least n bits (in fact, you need approximately $\sum_{j=1}^{n}\left(\log_2(w_j)+\log_2(v_j)\right)$ bits). So the claim that the algorithm is exponential implies that n is a logarithmic function of b (*i.e.*, $n \sim \log2(b)$).

Viewing the situation in a slightly different manner, note that the number of bits needed to represent the input data for the 0-1 knapsack problem is $L = \sum_{j=1}^{n}\left(\log_2(w_j)+\log_2(v_j)\right)+\log_2(b) \approx n+\log_2(b)$. As b gets large, we ask: Is there a polynomial of the form Lp for some finite p such that $b \leq Lp$? Looking at the ratio b/Lp and taking the limit we have

$$\lim_{b\to\infty}\frac{b}{(n+\log_2 b)^p}=\lim_{b\to\infty}\frac{b}{p!}=\infty$$

where the second ratio is derived from L'Hospital's rule. Hence, b cannot be bounded by a polynomial function of the input data. This result implies that the DP algorithm of requires and exponential number of steps with respect to the input data in the worst case.

1.12.5 NP-class of Solvable Problems

Deterministic algorithms are where no step is random. If the program/algorithm chooses a step non-deterministically (by some extraneous influence to the algorithm) such that it is always the *right choice*, then such an algorithm is called non-deterministic algorithm.

Example

Suppose a 0-1 KS backtrack-algorithm always knows which object to pick up next in order to find an optimal profit-making knapsack. If one has a *polynomial deterministic algorithm* for a problem (*e.g.*, the sorting problem), then the problem is called a **P-class** problem. And, the set of all such problems constitute the ***P-class.*** If one has a *polynomial non-deterministic algorithm* for a problem, then the problem is called a **NP-class** problem. And, the set of all such problems constitute the *NP-class.*

It is impossible to check for a problem's being in NP-class this way, because non-deterministic algorithms are impossible to develop, as defined above. So, how can one check for polynomial complexity of such non-existent algorithms?

However, an equivalent way of developing non-deterministic polynomial algorithm is: when a solution to the problem is provided (as if someone knows what the solution could be!), then that *proposed solution* is checked by that algorithm in *polynomial* time. Such a *proposed solution* is called a certificate to the input problem-instance.

For example: in a KSD problem, given a certificate *knapsack content* check if the total profit is $\geq p$ or not.

Complexity: calculation of total profit of the given knapsack content is worst case $O(n)$, for n objects.

For example: for a Hamiltonian circuit problem instance (*find a cycle in a graph over all nodes but without any node being repeated in the cycle*), a *path* is given as a certificate.

An algorithm would go over the certificate path and check if the first and the last nodes are *same*, and the rest of the path is *simple* (no node is covered twice), then it will check if all nodes in the graph are *covered* in the path, and then, check if all the *arcs* in the path do actually from the input graph. This takes polynomial time with respect to N and E. Hence HC is a NP-class problem.

Two important points to note:

1. NP-class problems are sub-set of the *Solvable* problems.
2. P-class problems are sub-set of NP-class problems (because if you have a deterministic algorithm to solve any problem instance, then that algorithm can be used to check any certificate in polynomial time also).

Example: non-HC problem (is there *no HC* in a given input graph) does not have any polynomial non-deterministic algorithm.

If a problem is in NP-class its complement (negative problem as the non-HC problem is) is in *Co-NP*.

All Co-NP problems constitute the Co-NP class of problems.

P-class is a subset of the intersection of NP and Co-NP sets.

From a reverse point of view, we would like to find an example problem that is in the NP-class and whose information-theoretic lower bound is exponential. Then, we would at least know that P is a proper subset of NP. We do not yet have any such example either. All we know now is that $\mathbf{P} \subseteq \mathbf{NP}$.

Question remains: (1) $P \subset NP$? or, $P \neq NP$? [One should find counter-example(s)]

Or, (2) $NP \subseteq P$, so that $P = NP$? [One should prove the theorem]

NP-completeness

Apparently some NP problems are "harder" in a relative sense than the other problems. If you can write polynomial algorithm for any one problem in this 'hard' group, then it can be shown that every NP-class problem will have a polynomial algorithm. This group of problems is called *NP-complete* problems. We will explain polynomial transformations first and then come back to the issue of NP-completeness.

1.12.6 Polynomial Transformation

Some algorithms which take a decision problem X (or rather ANY instance of the problem of type X), and output a corresponding instance of the decision problem of type Y, in such a way that if the input has answer *True*, then the output (of type *Y*) is also *True* and vice versa. [Problem ≡ Input; & Solution ≡ Algorithm that takes any input and produces correct output.]

For example, you can write a problem transformation algorithm from *3-SAT* problem to *3D-Matching* problem. Note that the problem transformations are directed.

When a problem transformation algorithm is polynomial-time we call it a **polynomial transformation.**

Existence of a polynomial transformation algorithm has a great **significance** for the complexity issues.

Suppose you have (1) a poly-transformation A_{xy} exists from a (source) problem X to another (target) problem Y, and (2) Y has a poly algorithm P_y, then you can solve any instance of the source problem X polynomially, by the following method.

Just transform any instance of X into another instance of Y first using A_{xy}, and then use Y's poly-algorithm P_y. Both of these steps are polynomial, and the output (*T/F*) from Y's algorithm is valid for the source instance (of X) as well. Hence, the *True/False* answer for the original instance of P_y (A_{xy} (X)) will be obtained in poly-time. This constitutes an indirect poly-algorithm for X, thus making X also belonging to the P-class. Note, $|A_{xy}(X)|$ is polynomial with respect to $|X|$.

1.12.7 Cook's Theorem

Cook modeled all NP-problems (an infinite set) to an abstract *Turing machine*. Then he developed a poly-transformation from this machine (*i.e.*, all NP-class problems) to a particular decision problem, namely, the *Boolean Satisfiability* (*SAT*) problem.

Significance of Cook's theorem: if one can find a poly-algorithm for SAT, then by using Cook's poly-transformation one can solve all NP-class problems in poly-time (consequently, P-class = NP-class would be proved).

SAT is the historically first identified *NP-hard problem*!

Further significance of Cook's theorem: if you find a poly-transformation from SAT to another problem Z, then Z becomes another NP-hard problem. That is, if anyone finds a poly algorithm for Z, then by using your poly-transformation from SAT-to-Z, anyone will be able to solve any SAT problem-instance in poly-time, and hence would be able to solve all NP-class problems in poly-time (by Cook's theorem).

These problems, which have a chain of poly-transformation from SAT, are called **NP-hard problems.**

If an NP-hard problem also belongs to the NP-class it is called an NP-complete problem, and the group of such problems are called **NP-complete problems.**

Significance of NP-hard Problems

As stated before, if one finds any poly-algorithm for any NP-hard problem, then

We would be able to write polynomial algorithm for each of NP-class problems, or

NP-class = P-class will be proved (in other words, poly-algorithms would be found for all NP-class problems).

Unfortunately, **neither** anyone could find any poly algorithm for any NP-hard problem (which would signify that P-class = NP-class); **nor** anyone could prove an exponential information-theoretic bound for any NP-complete problem, say problem L (which would signify that L is in NP but not in P, or in other words that would prove that P-class $\subset$ NP-class). The NP-hard problems are the best candidate for finding such counter-examples.

As a result, when we say a problem X (say, the KSD problem) is NP-complete, all we mean is that IF one finds a poly-alg for X, THEN all the NP-class problems would have poly algorithm.

We also mean, that it is UNLIKELY that there would be any poly-algorithm found for X.

P $\neq$ NP is a mathematical **conjecture** currently. YET to be proved as a theorem, (*see above though*). Based on this conjecture many new results about the complexity-structure of computational problems have been obtained.

For example, all problems, which need polynomial memory-space (rather than time) form PSPACE problems. Poly-time algorithm will not need more than poly-space, but the reverse may not be necessarily true. Answer to this question is not known. There exists a chain of poly-transformations from all PSPACE problems to the elements of a subset of it. That subset is called *PSPACE-complete*. PSPACE-complete problems may lie outside NP-class, and may be harder than the NP-complete problems. Example: *Quantified Boolean Formula* (first-order logic) is PSPACE-complete.

(SAT to 3-SAT)

SAT Problem, or Boolean Satisfiability Problem

(1) A set of Boolean variables, U = {a, b, c}. (2) A (conjunctive) set of *clauses* each consisting of a (disjunctive) set of *literals* (a variable or its negation), C = {{a, b}, {~a, ~b}, {a, c}, {~a, ~c}}. (3) Question: does there exist any set of assignments for the variables such that the clause is True. A clause is True if the assignment makes at least one literal true in the clause.

Output: An answer: *yes/no* (decision problem).

Answer to the above problem instance: *Yes* (for a=T, b=F, and c=F).

A backtracking algorithm for solving SAT is easy to devise. *SAT is in NP-class*: given a certificate truth assignment you can trivially check if it leads to the "Yes" answer or not, and you can do that in linear time O(*n*, *m*), for *n* number of variables, *m* number of clauses.

SAT is NP-hard: **Cook's theorem**.

k-SAT problem: Number of literals in each clause in the SAT problem is limited to *k*, where *k* is obviously an integer. It is not unreasonable to expect that *k-SAT*, for a constant integer *k*, to be possibly a P-class problem! K-SAT is restricted version of SAT. *2-SAT problem is* indeed *P-class*: Davis-Putnam algorithm can find an assignment of variables in polynomial time, always (for all problem instances), when *k*<=2. *BUT,*

3-SAT is NP-hard

The chosen source problem, which is already proved to be NP-hard, is the SAT problem itself. Transformation from **ANY** instance of SAT to a derived instance of 3-SAT is given below. Classify SAT clauses into 4 groups:

- 1-clauses
- 2-clauses
- 3-clauses
- p-clauses with p>3 literals in these clauses

We will:

1. show transformations (to the corresponding 3-clauses) for each of these 4 types of clauses;
2. prove that the transformations are correct, *i.e.*, truth preserving; and then
3. we will discuss the polynomial nature of the aggregate-transformation at the end.

1-clauses

Say, {u}, where u is a literal (not just a variable, but a variable or its negation). The corresponding 3-clauses (for the target 3-SAT problem) will need two additional variables, say, z1, z2. The derived 3-clauses are: {u, z1, z2}, {u, ~z1, z2}, {u, z1, ~z2}, {u, ~z1, ~z2}.

All these 3-clauses can be True only if *u* is True, and vice versa (hence this is a correct transformation).

2-caluses:

Say, {u1, u2}.

The corresponding 3-clauses (for the target 3-SAT problem) will need one additional variable, say, z.

The derived 3-clauses are: {u1, u2, z}, {u1, u2, ~z}.

Both the 3-clauses can be *True* only if either of u1 or u2 is *T* (*i.e.*, the source clause in the source SAT problem instance), and vice versa. Hence this is a correct transformation.

3-caluses

Say, {u1, u2, u3}.

Just copy them to the target problem instance, they are already 3-clauses.

p-caluses

Say, $\{u_1, u_2, \ldots, u_{k-1}, u_k, u_{k+1}, \ldots, u_p\}$, where $1 \le k \le p$.

Create the following 3-clauses (for the target 3-SAT problem):

$\{u_1, u_2, z_1\}, \{\sim z_1, u_3, z_2\}, \{\sim z_2, u_4, z_3\}, \ldots, \{\sim z_{k-3}, u_{k-1}, z_{k-2}\}, \{\sim z_{k-2}, u_k, z_{k-1}\}, \{\sim z_{k-1}, u_{k+1}, z_k\}, \ldots, \{\sim z_{p-3}, u_{p-1}, u_p\}$.

This needs creation of additional (p-3) variables: $z_1, z_2, \ldots, z_{p-3}$. In order for the *source (SAT) p-clause to be True,* at least one of the literals $u_1, \ldots, u_p$ has to be true. Without any loss of generality assume that the *True* literal to be u_k. We do not care if more literals are *True,* assume all the rest to be *False*. Then the following assignment of new variables (z's) (along with u_k=T) will make all the derived 3-clauses *True*:

z_1=T, z_2=T, ..., z_{k-3}=T, z_{k-2}=T, z_{k-1}=F, ..., z_{p-3}=F.

Hence, if the source *p*-clause has a *True* assignment, so do all the derived corresponding 3-clauses. Note that the construction algorithm does not care about which literal is really *True*.

Assume the source p-clause to be False, i.e., none of the literals $u_1, \ldots, u_p$ is *True*. Try any set of assignments for the new variables (z's), at least one clause will always remain *False,* you cannot make all of them *True* just by assigning the new variables.

Try one such assignment: Suppose, z_1=T (in order to make the first derived 3-clause *True*), then z_2 should be = T, and then z_3=T, ..., z_{p-3}=T (in order to make the last-but-one derived 3-clause *True*), but that makes the LAST derived 3-clause is *False*.

Try any other combination of assignments, you will always run into the same problem – *at least one derived clause will become false,* because that is how the derived 3-clauses are constructed. Hence, the transformation-scheme for the p-clauses is *truth-preserving,* or the algorithm is *correct*.

Complexity of the Transformation Algorithm

If a number of source (SAT) 1-clauses is k_1, 2-clauses is k_2, 3-clauses is k_3, and p-clauses are k_p (depending on the values of p), then, number of variables created in target 3-SAT problem instance would be $= n + 2k_1 + k_2 + \sum_p [(p-3)k_p]$.

The number of derived 3-clauses would be = $4k_1 + 2k_2 + k_3 + \sum_p [(p\text{-}2)k_p]$.

The total number of steps in creating the new variables and the new 3-clauses are the sum of these two polynomials, hence a *polynomial* with respect to the input problem size (#variables and #clauses in the source SAT problem instance).

Thus, we have a correct (truth preserving) & polynomial problem transformation algorithm from any SAT problem instance to a corresponding 3-SAT problem instance => 3-SAT is NP-hard (given the fact that SAT is NP-hard).

Since we have shown before that SAT is in NP-class, 3-SAT is obviously in NP-class. This concludes the proof that **3-SAT is NP-complete**.

Interval Temporal Reasoning Problem is NP-hard

Table 1.5: Basic Relations between Pairs of Intervals (B)

Binary operator	*Relation type*	*Example*	*Representation*
B	before	A(p)B	
~b	before inverse	A(~p)B	
O	overlap	A(o)B	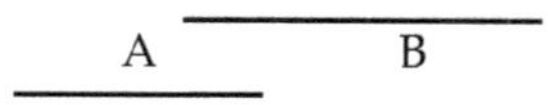
~o	overlap inverse	A(~o)B	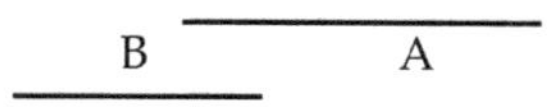
D	during	A(d)B	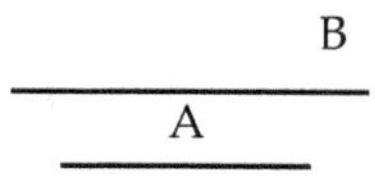
~d	during inverse	A(~d)B	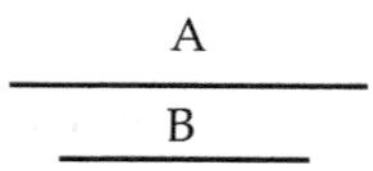
M	meet	A(m)B	
~m	meet inverse	A(~m)B	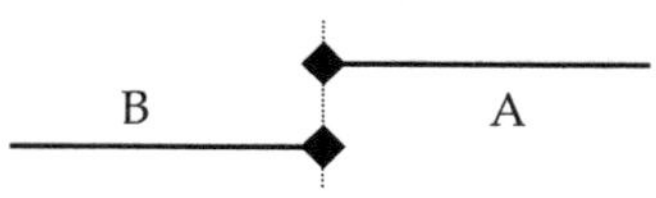
S	start	A(s)B	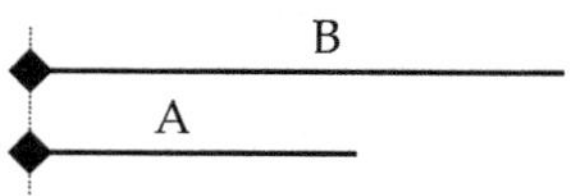
~s	start inverse	A(~s)B	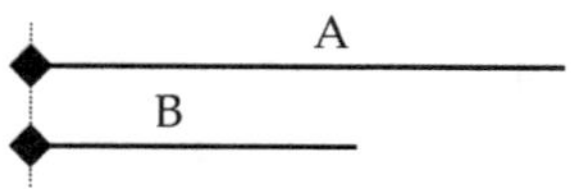

Binary operator	Relation type	Example	Representation
F	finish	A(f)B	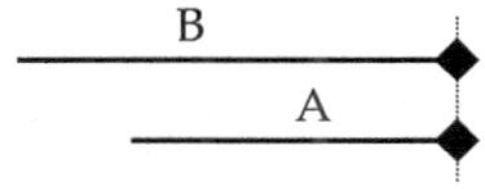
~f	finish inverse	A(~f)B	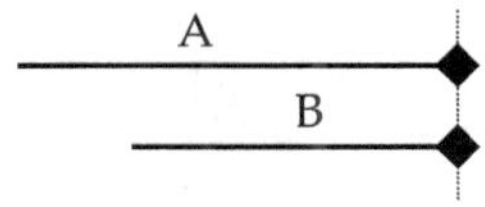
Eq	equal	B(eq)A	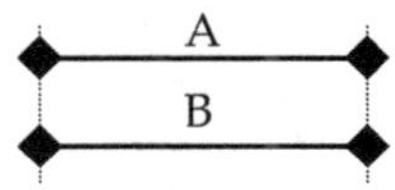

QTCN: A *qualitative temporal-constraint network* (QTCN) is a graph $G=(V, E)$, where each node is an interval, and each directed labeled edge $(v_1\ (R)\ v_{2'})\in E$ represents disjunctive constraint R from v_1 to $v_2 \in V$, where $R \in 2^B$.

ITR Problem: Given a QTCN does there exist an assignment for each node on the time-line such that all the constraints are satisfied.

Example: QTCN[V = {i1, i2, i3}, E = {(i1 (p|o) i2), (i1 (m) i3), (i2 (o|a) i3)}]. Answer to ITR: Yes. Draw i1, i2 and i3 on timeline in such a way that i1 overlaps i2 and meets i3, while i2 overlaps i3.

ITR is NP-hard: transform arbitrary 3-SAT problem to the corresponding ITR problem.

Let, the input 3-SAT problem is C = {(li1, li2, li3) | i = 1, …, m}, i indicates the clause, lik indicates k-th literal in the i-th clause.

Create an interval called *spl*. Create an interval Iij for each literal lij. Imagine left of *spl* will have the True literals (when the 3-SAT is assigned) and right of *spl* will contain the False literals. All three literals of a clause can be on the left of *spl* but not more than two can be on the right, when the input clause is satisfied by the assignment. This is the trick in creating interval relations.

First, fix the intervals so that they can be on either side of *spl* but not on both the side, with a constrained end point:

For all i=1,..,m and j=1, 2, 3

Iij (m|~m) spl.

Then the other constraint, for each clause i = 1..m,

Ii1 (f | ~f | s | p | ~p) Ii2

Ii2 (f | ~f | s | p | ~p) Ii3

Ii3 (f | ~f | s | p | ~p) Ii1

Finally, we want to make sure that when two literals appear with opposite sign in different clauses, their corresponding intervals go on different sides of the interval *spl*.

For each lij = ~lgh, a constraint

Iij (p | ~p) Igh

This construction ensures that if the source 3-SAT is satisfiable one can easily satisfy the generated ITR. However, if the source 3-SAT is unsatisfiable (at least on clause, say, Ck is unsatisfiable by any given assignment), then the corresponding intervals in the ITR cannot be assigned as all the three intervals for that clause will try to be on the False side, which is impossible as per the construction ('s' relation's inverse is absent thus, creating a cycle over the intervals Ik1, Ik2, and Ik3).

3D Matching is NP-complete

2DM Problem: Two finite sets of individuals of equal cardinality: M = {m1, m2, … mk}, F = {f1, f2, …, fk}. A set of acceptable pairs of individuals from the two groups: A = {(fi, mj), 1 ≤ i,j ≤ k} = {(f3, m1), (f5, m1), (f3, m2), (f9, m2), …}.

Question: Does there exist a matching set of tuples A′ within A (subset of A) such that every individual is in a tuple in A′ once and only once. The cardinality of A′ will be k, if it does exist.

3DM Problem: 3 sets (or "types") of individuals, X, Y, W, with same cardinality. Each element in A is a 3-tuple, A = $((x_i, y_j, w_l), 1 \le i, j, l \le k\}$ from those three sets.

Question: similar as above (does there exist a matching subset A′ within A).

3DM is NP-class: Given a matching set A′ one has to check: (1) Every individual is covered or not; (2) Any individual is covered twice or not. It can be done in polynomial time :: check in a 3D coordinate system.

3DM is NP-hard: Polynomial transformation from 3-SAT to 3DM

Source 3-SAT Problem: Variables U=$\{u_1, u_2, \ldots, u_n\}$, 3-Clauses C=$\{c_1, c_2, \ldots, c_m\}$.

We have to construct 3 sets of individuals X, Y, and W; and a set of triplets A from them. *Truthfulness property needs to be preserved*. Transformation algorithm must be polynomial with respect to number of variables and clauses in the input 3-SAT problem.

3 types of triplets in A: "truth-setting and fan-out," "satisfaction testing" (of source 3-SAT), and "garbage collection" (to make sure everybody is covered in the target 3DM instance).

Truth-Setting + Fan-out Components of A

In each clause each variable may appear only once (as Positive or as Negative)

For each variable u_i $(1 \le i \le n)$ and each clause c_j $(1 \le j \le m)$: two sets of triplets in A:

$T_{pos\_i} = \{ (\underline{u}_i[j], a_i[j], b_i[j]) : 1 \leq j \leq m\}$

$T_{neg\_i} = \{ (u_i[j], a_i[j+1], b_i[j]) : 1 \leq j < m\} + \{ (u_i[m], a_i[1], b_i[m])\}$

a's are in X (m), and b's are in Y (m), u's and $\underline{u}$'s are in W (2m).

For every variable u_i, the matching will have to pick up exactly m triplets from A (out of 2m), for A', depending on whether u_i is assigned *True* (T_{pos_i}) or *False* (T_{neg_i}).

Total number of triplets = 2mn

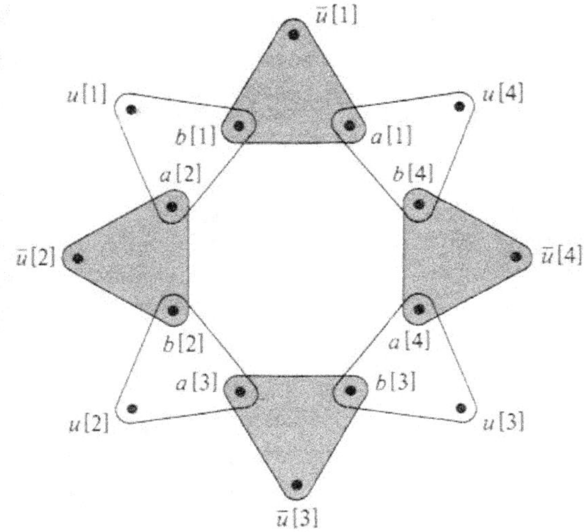

Figure 1.1: Truth Setting Component T.

Satisfaction-Testing Components of A

A triplet for each literal in each clause (**3m in number**), this is where we need 3-SAT (or a k-SAT, for a fixed integer k, as a source problem, as opposed to the general SAT problem - in the later case this number may not remain polynomial).

For each clause c_j three triplets (corresponding to three literals in it):

Cj = { either $(\underline{u}_i[j], s_a[j], s_b[j])$ if $\sim u_i$ is in c_j, or $(u_i[j], s_a[j], s_b[j])$ if u_i is in c_j}.

Cj's are components within A.

When the input/source 3-SAT has a satisfying assignment, the variable u (or $\underline{u}$) corresponding to the *True* literals will not have match as per previous construction (with a's and b's), and hence this construction Cj will get them matched with corresponding s[]'s. On the other hand if a clause has all its literals *False*, every u (or $\underline{u}$) in the corresponding Cj is matched already in the previous construction, but if you do not pick up from Cj, then the corresponding s's remain unmatched.

Garbage Collecting Components of A

To get those individuals who are still not matched in A: not all variables occur in every clause, but the corresponding u's are created anyway

Two sets of triplets in A for each variable u_i and each clause cj:

G = {($\underline{u}_i$[j], g_a[k], g_b[k]), (u_i[j], g_a[k], g_b[k]) : 1 <= k <= m(n-1), 1 <= i <= n, 1 <= j <= m}.

One set per k, k runs up to m(n-1). **Total triplets (2mn).m(n-1).**

For the u's (or $\underline{u}$'s) that are not in A' (or, not in A???).

Sets of individuals:

W = {$\underline{u}_i$[j], u_i[j] : 1≤i≤n, 1≤j≤m}, **2mn** in number

X = Aa + S1 + G1

Aa = {a_i[j] : 1≤i≤n, 1≤j≤m},	mn
Sa = {s_a[j] : 1≤j≤m},	m
Ga = {g_a[k] : 1≤k≤m(n-1)},	m(n-1)
	Total=**2mn**

Y = Bb + S2 + G2

Bb = {b_i[j] : 1≤i≤n, 1≤j≤m},	mn
Sb = {s_b[j] : 1≤j≤m},	m
Gb = {g_b[k] : 1≤k≤m(n-1)},	m(n-1)
	Total=**2mn**

Polynomial Construction

#individuals + #triplets = 6mn + (2mn + 3m + 2(m^2)n(n-1))

Correctness

If C is not satisfiable A' cannot be constructed. If A' is constructed, then the triplets in A' corresponding to the s1's and s2's are the truth-setting components, make the corresponding literals $\underline{u}$ or u's from those triplets *True*, there is no way the same variable will get multiple conflicting assignments by this.

If C is satisfiable:

Say, there exists an assignment for each u_i (*True*, or *False*) which makes C *True*.

For constructing A':

Choose a literal from each clause that is *True* (one must exist), and find corresponding triplet from the Satisfaction-testing components, for marrying each corresponding s1 and s2. Depending on if u_i is *True* or *False* choose the set T_{pos_i} or T_{neg_i}, to marry off a's and b's. Pick up appropriate triplets from G to take care of g_a's and g_b's, for those u_1's and u_2's who are still unmarried. Thus, A′ can be constructed, from the source 3-SAT's variable assignments.

Example with SAT->3DM trans:

3-SAT: var: (u1, u2, u3), clause: {(u1, ~u2, u3)}

3DM:

Truth setting construction:

A triplets:		Unmatched individuals:
Tpos1=(~u11, a11, b11)	Tneg1=(u11, a11, b11)	W={~u11, u11, ~u21, u21, ~u31, u31}
Tpos2=(~u21, a21, b21)	Tneg2=(u21, a21, b21)	X={a11, a21, a31}
Tpos3=(~u31, a31, b31)	Tneg3=(u31, a31, b31)	Y={b11, b21, b31}

A′ match set for sat assignment (say, for u1=T, u2=T and u3=F) in source-3SAT problem:

{(~u11, a11, b11), (~u21, a21, b21), (u31, a31, b31)}

Unmatched individuals left after this:

W={u11, u21, ~u31}, X={none}, Y={none}

Satisfaction testing construction:

A triplets:	Unmatched individuals:
C1={(u11, sa1, sb1), (~u21, sa1, sb1), (u31, sa1, sb1)}	W={u11, u21, ~u31}
X={sa1, sa2},	Y={sb1, sb2}

A′ match set for truth setting of the clause, say, by u1=T:

{(u11, sa1, sb1)}

Unmatched individuals left after this:

W={u21, ~u31}, X={none}, Y={none}

Garbage collection/ create new individuals to match all:

A triplets:	Unmatched individuals:
G={(~u11, ga1, gb1), (u11, ga1, gb1),	W={u21, ~u31}
(~u21, ga1, gb1), (u21, ga1, gb1),	X={ga1, ga2}
(~u31, ga1, gb1), (u31, ga1, gb1),	Y={gb1, gb2}
(~u11, ga2, gb2), (u11, ga2, gb2),	
(~u21, ga2, gb2), (u21, ga2, gb2),	
(~u31, ga2, gb2), (u31, ga2, gb2)}	

1<=k<=1(3-1), 1<=i<=3, 1<=j<=1;

A′ match set for truth setting of the clause, say, by u1=T:

{(u21, ga1, gb1), (~u31, ga2, gb2)}

Unmatched individuals left after this:

W={None}, X={none}, Y={none}

You may like to try with a unsatisfiable source 3-SAT:

{u, v, w}, {u, v, ~w},{u, ~v, w},{u, ~v, ~w},

{~u, v, w}, {~u, v, ~w},{~u, ~v, w},{~u, ~v, ~w}.

Take u=v=w=T that makes the last clause *False*. Try to create triplets in A for that clause and see why you cannot create match for corresponding sa's and sb's created in the Truth-setting component of the construction.

Reasoning with Cardinal-Directions Algebra

Input: A set of points V, and a set E of binary relations between some pairs of points in V, $v_i R_{ij} v_j$ where R_{ij} is a disjunctive subset of the set of nine basic relations {Eq, East, ...} in Cardinal directions-calculus.

Question: Can the points in V be located in a real space of two-dimensions.

Example input (1): V={v1, v2, v3, v4}, and E = {(v2 (Northeast, North) v1), (v3 (Northeast) v2), (v3 (West, Southwest, South) v1), (v4 (North) v3)}. Answer: No satisfying placement of v1, v2, v3 and v4 exists following these constraints.

Example input (2): V={v1, v2, v3, v4}, and E = {(v2 (Northeast, North) v1), (v3 (Northeast) v2), (v3 (Northeast) v1), (v4 (North) v3)}. Answer: Yes. v1 = (0,0), v2=(1,1), v3=(3,3), and v4=(3,4) is such a satisfying assignment.

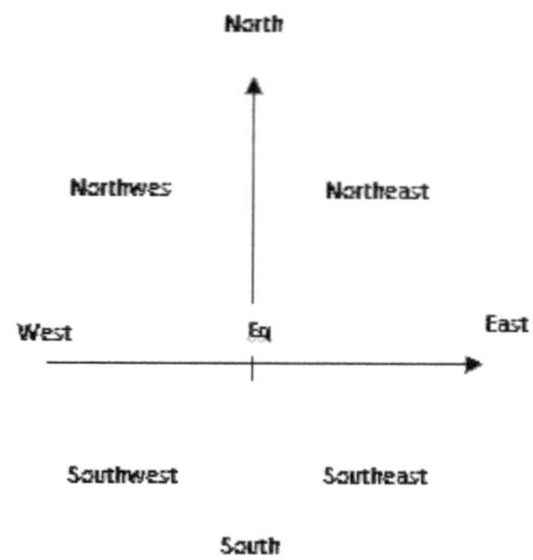

Figyre 1.2: 2D-Cardinal Directions Calculus.

Theorem: reasoning with 2D-Cardinal directions algebra is NP-hard.

Proof: By constructing a Cardinal-directions algebra problem instance from an arbitrary 3-SAT problem instance using a set of clauses like $C_i = \{ l_{i,1} \mid l_{i,2} \mid l_{i,3} \}$ where:

(1) For every literal l_{ij} create two points P_{ij} and R_{ij} such that P_{ij} {nw, n, ne, e, se} R_{ij}, and (2) for every clause C_i we have P_{i1} {sw, w, nw} R_{i2} and P_{i2} {sw, w, nw} R_{i3} and P_{i3} {sw, w, nw} R_{i1}. Also, (3) for every literal l_{ij} that has a complementary literal l_{gh} we have two relations between their corresponding points: P_{ij} {sw, s, se} R_{gh} and P_{gh} {sw, s, se} R_{ij}.

Writing P as (p, q) and R as (r, s) we get three relations on x-axis and y-axis,

(1) $(p_{ij} > r_{ij}) \mid (q_{ij} > s_{ij})$,

(2) $(p_{i1} < r_{i3})$ & $(p_{i2} < r_{i1})$ & $(p_{i3} < r_{i2})$,

(3) If l_{ij} and l_{gh} are complementary literals, $(q_{gh} < s_{ij})$ & $(q_{ij} < s_{gh})$

e express $(p_{ij} > r_{ij})$ as false whenever (if and only if) any literal l_{ij} is *true* in clause C_i.

So, for the six points corresponding to any unsatisfiable clause where all literals are *false*, generates three relations for their x-coordinates, $(p_{i1} > r_{i1})$ & $(p_{i2} > r_{i2})$ & $(p_{i3} > r_{i3})$ by construction (1). In addition by construction (2) we get three more relations $(p_{i1} < r_{i3})$ & $(p_{i2} < r_{i1})$ & $(p_{i3} < r_{i2})$. You cannot assign these six points on the x-axis satisfying all these six relations. Hence, if any clause is *false* (by any truth assignment of literals), then the corresponding Cardinal-directions does not have a solution as the six points cannot be assigned in space.

On the other hand suppose there exists a truth assignment that satisfies all clauses. So for every clause C_i there is at least one literal l_{ij} *true*, or $(p_{ij} > r_{ij})$ is false, or $\mathbf{(q_{ij} > s_{ij})}$ is true (by construction (1)). Then, a complementary literal l_{gh} in clause C_g will create the following relations: $\mathbf{(q_{gh} < s_{ij})}$ **&** $\mathbf{(q_{ij} < s_{gh})}$**.** The three relations are not consistent with $\mathbf{(q_{gh} > s_{gh})}$ that should be true if and only if l_{gh} is also *true*. Hence, l_{ij} and l_{gh} cannot be *true* at the same time, thus, prohibiting any inconsistency for the truth assignment of literals across the clauses. In other words, the cross-linking of the points across the corresponding literals will not prohibit from their being put in the space. Note that as soon as a literal is *true* in a clause the corresponding six points can be assigned as per constructions 1 and 2, only construction 3 might have created problem via the complementary literals.

The number of steps in the construction algorithm involves: constructing 6 points per clause, 6 relations per clause from construction (1), 3 relations per clause from (2), and at the most 3 relations per pair of clauses. Total number is polynomial with respect to the numbers of variables and clauses. Hence, the above construction is a polynomial transformation from 3-SAT problem to the 2D-Cardinal algebra problem.

Other Models of Compuation

NP-hardness came as a big shock to the progress of computing. Many of the important and interesting problems turn out to be NP-hard!

Areas where problems are not NP-hard see success in the market place, *e.g.*, most problems in data-management or linear programming. Other areas where problems are NP-hard either bypass (see discussion above on how to tackle NP-hard problems) or simply continue to exist as a research topic for a long time (*e.g.*, artificial intelligence, or software engineering).

Note that the model of computation in which some problems are NP-hard is the Turing/Church (TM) model of computation.

The first hope was raised with **parallel computing**. Alas, that turned out to be not more powerful than the present model!

Significance: even if infinite number of CPU's are provided and communication time is 0 between them, the NP-hard problems may not still have any distributed poly-algorithms.

Artificial Neural Network provides a different model of computation. It has the same power as the Turing Machine, but it solves some pattern recognition problems very fast.

An esoteric approach is the **DNA computing**. They solve some graph-theoretic problems [*e.g.* TSP] in a very small number of steps (but with a huge time per step with the current bio-technology). But DNA computing is proved to be as powerful as Turing Machine (*i.e.*, NP-h remains NP-h even with DNA computing). Researchers are trying to develop computers out of DNA's.

Another limitation to the conventional computing is surfacing from the *hardware*: circuit densities on chips are becoming too large: avoiding cross-feeds between circuit-elements will become impossible soon, because of the "quantum-tunneling" effects.

A new technology shaping up on the horizon is in the form of **Quantum computing**. Multiple data (and instructions) can be "superimposed" in the same quantum-register. When needed, a quantum operator will "extract" the required output. An algorithm becomes a Quantum operator in this paradigm.

Physical technological hurdles are being overcome gradually. It seems we have been doing quantum computing unknowingly for the last few decades in the form of NMR (nuclear magnetic resonance, as in MRI in medicine), where atomic nuclei provide the Quantum bits (Q-bits).

Quantum computing (QC) is a different model than TM, has in-built non-determinism. But, apparently NP-hardness remain NP-hard their too *(find out a relevant paper for a free lunch from me)!* However, speed, space complexity, and power consumption of QC may be much lower than those of the electronic computers.

Algorithms for QC are quite different than those for conventional computers (based on Turing machine). QC may also provide a better model for security, which attracts cryptographers and the communications industry. QC has also generated topic called *Spintronics* where electrons' spin is used instead of charge for building digital circuitry.

UNIT 2

Linear Programming

2.1 Introduction

In mathematics, linear programming (LP) is a technique for optimization of a linear objective function, subject to linear equality and linear inequality constraints. Informally, linear programming determines the way to achieve the best outcome (such as maximum profit or lowest cost) in a given mathematical model and given some list of requirements represented as linear equations.

More formally, given a polytope (for example, a polygon or a polyhedron), and a real-valued affine function

$$f(x_1, x_2, \ldots, x_n) = c_1x_1 + c_2x_2 + \cdots + c_nx_n + d$$

defined on this polytope, a linear programming method will find a point in the polytope where this function has the smallest (or largest) value. Such points may not exist, but if they do, searching through the polytope vertices is guaranteed to find at least one of them.

Linear programs are problems that can be expressed in canonical form:

Maximize $c^T x$

Subject to $Ax \leq b$.

X represents the vector of variables (to be determined), while c and b are vectors of (known) coefficients and A is a (known) matrix of coefficients. The expression to be maximized or minimized is called the objective function ($c^T x$ in this case). The equations $Ax \leq b$ are the constraints which specify a convex polyhedron over which the objective function is to be optimized.

Linear programming can be applied to various fields of study. Most extensively it is used in business and economic situations, but can also be utilized for some engineering problems. Some industries that use linear programming models include transportation, energy, telecommunications, and manufacturing. It has proved useful in modeling diverse types of problems in planning, routing, scheduling, assignment, and design.

2.1.1 Theory

Geometrically, the linear constraints define a convex polyhedron, which is called the feasible region. Since the objective function is also linear, hence a convex function, all local optima are automatically global optima (by the KKT theorem). The linearity of the objective function also implies that the set of optimal solutions is the convex hull of a finite set of points — usually a single point.

There are two situations in which no optimal solution can be found. First, if the constraints contradict each other (for instance, $x \geq 2$ and $x \leq 1$) then the feasible region is empty and there can be no optimal solution, since there are no solutions at all. In this case, the LP is said to be *infeasible*.

Alternatively, the polyhedron can be unbounded in the direction of the objective function (for example: maximize $x_1 + 3\,x_2$ subject to $x_1 \geq 0$, $x_2 \geq 0$, $x_1 + x_2 \geq 10$), in which case there is no optimal solution since solutions with arbitrarily high values of the objective function can be constructed.

Barring these two pathological conditions (which are often ruled out by resource constraints integral to the problem being represented, as above), the optimum is always attained at a vertex of the polyhedron. However, the optimum is not necessarily unique: it is possible to have a set of optimal solutions covering an edge or face of the polyhedron, or even the entire polyhedron (This last situation would occur if the objective function were constant).

2.2 Uses

Linear programming is a considerable field of optimization for several reasons. Many practical problems in operations research can be expressed as linear programming problems. Certain special cases of linear programming, such as network flow problems and multicommodity flow problems are considered important enough to have generated much research on specialized algorithms for their solution. A number of algorithms for other types of optimization problems work by solving LP problems as sub-problems. Historically, ideas from linear programming have inspired many of the central concepts of optimization theory, such as duality, decomposition, and the importance of convexity and its generalizations. Likewise, linear programming is heavily used in microeconomics and company management, such as planning, production, transportation, technology and other issues. Although the modern management issues are ever-changing, most companies would like to maximize profits or minimize costs with limited resources. Therefore, many issues can boil down to linear programming problems.

2.3 Standard Form

Standard form is the usual and most intuitive form of describing a linear programming problem. It consists of the following three parts:

- A **linear function to be maximized**

  e.g., maximize $c_1 x_1 + c_2 x_2$

- **Problem constraints** of the following form

  e.g.,

  $a_{11}x_1 + a_{12}x_{12} \le b_1$

  $a_{21}x_1 + a_{22}x_2 \le b_2$

  $a_{31}x_1 + a_{32}x_2 \le b_3$

- **Non-negative variables**

  e.g. $x_1 \ge 0$

  $x_2 \ge 0$

The problem is usually expressed in *matrix form*, and then becomes:

maximize $c^T x$

subject to $Ax \le b,\ x \ge 0$

Other forms, such as minimization problems, problems with constraints on alternative forms, as well as problems involving negative variables can always be rewritten into an equivalent problem in standard form.

2.4 Examples

Example 1

Suppose that a farmer has a piece of farm land, say *A* square kilometres large, to be planted with either wheat or barley or some combination of the two. The farmer has a limited permissible amount *F* of fertilizer and *P* of insecticide which can be used, each of which is required in different amounts per unit area for wheat (F_1, P_1) and barley (F_2, P_2). Let S_1 be the selling price of wheat, and S_2 the price of barley. If we denote the area planted with wheat and barley by x_1 and x_2 respectively, then the optimal number of square kilometres to plant with wheat vs barley can be expressed as a linear programming problem:

maximize $S_1x_1 + S_2x_2$		(maximize the revenue — revenue is the "objective function")
subject to	$x_1 + x_2 \le A$	(limit on total area)
	$F_1x_1 + F_2x_2 \le F$	(limit on fertilizer)

	$P_1x_1 + p_2x_2 \geq 0$	(limit on insecticide)
	$x_1 \geq 0, x_2 \geq 0$	(cannot plant a negative area)

Which in matrix form becomes:

maximize $\begin{bmatrix} S_1 S_2 \end{bmatrix} \begin{bmatrix} x_1 \\ x_2 \end{bmatrix}$

subject to $\begin{bmatrix} 1 & 1 \\ F_2 & F_2 \\ P_1 & P_2 \end{bmatrix} \begin{bmatrix} x_1 \\ x_2 \end{bmatrix} \leq \begin{bmatrix} A \\ F \\ P \end{bmatrix}, \begin{bmatrix} x_1 \\ x_2 \end{bmatrix} \geq 0$

2.4.1 Augmented Form

Linear programming problems must be converted into *augmented form* before being solved by the simplex algorithm. This form introduces non-negative *slack variables* to replace inequalities with equalities in the constraints. The problem can then be written in the following form:

Maximize Z in:

$$\begin{bmatrix} 1 & -c^T & 0 \\ 0 & A & I \end{bmatrix} \begin{bmatrix} Z \\ x \\ x_s \end{bmatrix} = \begin{bmatrix} 0 \\ b \end{bmatrix}$$

$$x, x_s \geq 0$$

where $\mathbf{X}_S$ are the newly introduced slack variables, and Z is the variable to be maximized.

Example 2

The example above becomes as follows when converted into augmented form:

maximize $S_1x_1 + S_2x_2$		(objective function)
subject to	$x_1 + x_2 + x_3 = A$	(augmented constraint)
	$F_1x_1 + F_2x_2 + x_4 = F$	(augmented constraint)
	$P_1x_1 + P_2x_2 + x_5 = P$	(augmented constraint)
	$x_1, x_2, x_3x_4x_5 \geq 0$	

where x_3, x_4, x_5 are (non-negative) slack variables.

Which in matrix form becomes:

Maximize Z in:

$$\begin{bmatrix} 1 & -S & -S_2 & 0 & 0 & 0 \\ 0 & 1 & 1 & 1 & 0 & 0 \\ 0 & F_1 & F_2 & 0 & 1 & 0 \\ 0 & P_1 & P_2 & 0 & 0 & 1 \end{bmatrix} \begin{bmatrix} Z \\ x_1 \\ x_2 \\ x_3 \\ x_4 \\ x_5 \end{bmatrix} = \begin{bmatrix} 0 \\ A \\ F \\ P \end{bmatrix}, \begin{bmatrix} x_1 \\ x_2 \\ x_3 \\ x_4 \\ x_5 \end{bmatrix} \geq 0$$

2.4.2 Duality

Every linear programming problem, referred to as a primal problem, can be converted into a dual problem, which provides an upper bound to the optimal value of the primal problem. In matrix form, we can express the *primal problem* as:

maximize $c^T x$

subject to $Ax \leq b, x \geq 0$

The corresponding *dual problem* is:

minimize $b^T y$

subject to $A^T y \geq c, y \geq 0$

where y is used instead of x as variable vector.

There are two ideas fundamental to duality theory. One is the fact that the dual of a dual linear program is the original primal linear program. Additionally, every feasible solution for a linear program gives a bound on the optimal value of the objective function of its dual. The weak duality theorem states that the objective function value of the dual at any feasible solution is always greater than or equal to the objective function value of the primal at any feasible solution. The strong duality theorem states that if the primal has an optimal solution, x*, then the dual also has an optimal solution, y*, such that $c^T x^* = b^T y^*$.

A linear program can also be unbounded or infeasible. Duality theory tells us that if the primal is unbounded then the dual is infeasible by the weak duality theorem. Likewise, if the dual is unbounded, then the primal must be infeasible. However, it is possible for both the dual and the primal to be infeasible.

Example 3

Revisit the above example of the farmer who may grow wheat and barley with the set provision of some A land, F fertilizer and P insecticide. Assume now that unit prices for each of these means of production (inputs) are set by a planning board. The planning board's job is to minimize the total cost of procuring the set amounts of inputs while providing the farmer with a floor on the unit price of each of his crops (outputs), S1 for wheat and S2 for barley. This corresponds to the following linear programming problem:

minimize	$Ay_A + Fy_F + Py_P$	(minimize the total cost of the means of production as the "objective function")
subject to	$y_A + F_1 y_F + P_1 y_P \ge S_1$	(the farmer must receive no less than S_1 for his wheat)
	$y_A + F_2 y_F + P_2 y_P \ge S_2$	(the farmer must receive no less than S_2 for his barley)
	$y_A \ge 0, y_F \ge 0, y_P \ge 0$	(prices cannot be negative)

Which in matrix form becomes:

$$\text{minimize} \begin{bmatrix} A & F & P \end{bmatrix} \begin{bmatrix} y_A \\ y_F \\ y_P \end{bmatrix}$$

$$\text{subject to} \begin{bmatrix} 1 & F_1 & P_1 \\ 1 & F_2 & P_2 \end{bmatrix} \begin{bmatrix} y_A \\ y_F \\ y_P \end{bmatrix} \ge \begin{bmatrix} S_1 \\ S_2 \end{bmatrix}, \begin{bmatrix} y_A \\ y_F \\ y_P \end{bmatrix} \ge 0$$

The primal problem deals with physical quantities. With all inputs available in limited quantities, and assuming the unit prices of all outputs is known, what quantities of outputs to produce so as to maximize total revenue? The dual problem deals with economic values. With floor guarantees on all output unit prices, and assuming the available quantity of all inputs is known, what input unit pricing scheme to set so as to minimize total expenditure?

To each variable in the primal space corresponds an inequality to satisfy in the dual space, both indexed by output type. To each inequality to satisfy in the primal space corresponds a variable in the dual space, both indexed by input type.

The coefficients which bound the inequalities in the primal space are used to compute the objective in the dual space, input quantities in this example. The coefficients used to compute the objective in the primal space bound the inequalities in the dual space, output unit prices in this example.

Both the primal and the dual problems make use of the same matrix. In the primal space, this matrix expresses the consumption of physical quantities of inputs necessary to produce set quantities of outputs. In the dual space, it expresses the creation of the economic values associated with the outputs from set input unit prices.

Since each inequality can be replaced by an equality and a slack variable, this means each primal variable corresponds to a dual slack variable, and each dual variable corresponds to a primal slack variable. This relation allows us to complementary slackness.

2.4.3 Special Case

A *packing LP* is a linear program of the form

maximize $c^T x$

subject to $Ax \leq b, x \geq 0$

such that the matrix A and the vectors *b* and *c* are non-negative.

The dual of a packing LP is a *covering LP*, a linear program of the form

minimize $b^T y$

subject to $A^T y \geq c, y \geq 0$

such that the matrix A and the vectors *b* and *c* are non-negative.

Example 4

Covering and packing LPs commonly arise as a linear programming relaxation of a combinatorial problem. For example, the LP relaxation of set packing problem, independent set problem, or matching is a packing LP. The LP relaxation of set cover problem, vertex cover problem, or dominating set problem is a covering LP.

Finding a fractional coloring of a graph is another example of a covering LP. In this case, there is one constraint for each vertex of the graph and one variable for each independent set of the graph.

2.5 Solutions to Linear Programming

2.5.1 Introduction

A Linear Programming problem is a special case of a Mathematical Programming problem. From an analytical perspective, a mathematical program tries to identify an extreme (*i.e.*, minimum or maximum) point of a function $f(x_1, x_2, \ldots, x_n)$, which furthermore satisfies a set of constraints, *e.g.*, $g(x_1, x_2, \ldots, x_n) \geq b$ Linear programming is the specialization of mathematical programming to the case where both, function *f* - to be called the objective function - and the problem constraints are linear.

Solution procedure used when a Linear Programming (LP) problem has two (or at most three) decision variables. The graphical method follows these steps:

1. Change inequalities to equalities.
2. Graph the equalities.
3. Identify the correct side for the original inequalities.
4. Then identify the feasible region, the area of Feasible Solution.
5. Determine the Contribution Margin (CM) or cost at each of the corner points (basic feasible solutions) of the feasible region.
6. Pick either the most profitable or least cost combination, which is an Optimal Solution.

2.5.2 Prototype LP Problem

Consider a company which produces two types of products P1 and P2 . Production of these products is supported by two workstations W1 and W2 , with each station visited by both product types. If workstation W1 is dedicated completely to the production of product type P1, it can process 40 units per day, while if it is dedicated to the production of product P2, it can process 60 units per day. Similarly, workstation W2 can produce daily 50 units of product P1 and 50 units of product P2 , assuming that it is dedicated completely to the production of the corresponding product. If the company's profit by disposing one unit of product P1 is \$200 and that of disposing one unit of P2 is \$400, and assuming that the company can dispose its entire production, how many units of each product should the company produce on a daily basis to maximize its profit?

Now the objective is to maximize the company's profit, which under the problem assumptions, is equivalent to maximizing the company's daily profit. Furthermore, we are going to maximize the company profit by adjusting the levels of the daily production for the two items P1 and P2 . Therefore, these daily production levels are the control/decision factors, the values of which we are called to determine. In the analytical formulation of the problem, the role of these factors is captured by modeling them as the problem decision variables:

- X1 := number of units of product P1 to be produced daily
- X2:= number of units of product P2 to be produced daily

In the light of the above discussion, the problem objective can be expressed analytically as:

$$\max f(X_1, X_2) := 200x_1 + 400x_2 \tag{1}$$

Equation above will be called the objective function of the problem, and the coefficients 200 and 400 which multiply the decision variables in it, will be called the objective function coefficients.

Furthermore, any decision regarding the daily production levels for items P1 and P2 in order to be realizable in the company's operation context must observe the production capacity of the two worksations W1 and W2 . Hence, our next step in the problem formulation seeks to introduce these technological constraints in it. Let's focus first on the constraint which expresses the finite production capacity of workstation W1. Regarding this constraint, we know that one day's work dedicated to the production of item P1 can result in 40 units of that item, while the same period dedicated to the production of item P2 will provide 60 units of it. Assuming that production of one unit of product type Pi, i=1, 2 requires a constant amount of processing time at workstation W1 , it follows that: $\tau_{11} = \frac{1}{40}$ and $\tau_{12} = \frac{1}{60}$. Under the further assumption that the combined production of both items has no side-effects, *i.e.*, does not impose any additional requirements for production capacity of workstation W1(*e.g.*, zero set-up times), the total capacity (in terms of time length) required for producing X1units of product P1and X2 units of product P2 is equal to $\frac{1}{40}X_1 + \frac{1}{60}X_2$. Hence, the technological constraint imposing the condition that our total daily processing requirements for workstation W1 should not exceed its production capacity, is analytically expressed by:

$$\frac{1}{40}X_1 + \frac{1}{60}X_2 \leq 1.0 \tag{2}$$

Notice that in Equation 2 time is measured in days.

Following the same line of reasoning (and under similar assumptions), the constraint expressing the finite processing capacity of workstation W_2 is given by:

$$\frac{1}{50}X_1 + \frac{1}{50}X_2 \leq 1.0 \tag{3}$$

Constraints 2 and 3 are known as the technological constraints of the problem. In particular, the coefficients of the variables $X_i, i = 1,2$ in them, $\frac{1}{\tau_{ji}}, j,i = 1,2$

$X_i, i = 1,2$, are known as the technological coefficients of the problem formulation, while the values on the right-hand-side of the two inequalities define the right-hand side (rhs) vector of the constraints.

Finally, to the above constraints we must add the requirement that any permissible value for variables $X_i, i = 1,2$ must be non-negative, *i.e.*,

$$x_i \geq 0 \quad i = 1,2 \tag{4}$$

since these values express production levels. These constraints are known as the variable sign restrictions.

Combining Equations 1 to 4, the analytical formulation of our problem is as follows:

$$\max f(x_1, x_2) := 200X_1 + 400X_2 \tag{5}$$

2.5.3 General LP Formulation

The general form for a Linear Programming problem is as follows:

Objective Function:

$$\max/\min f(X_1, X_2, \ldots, X_n) := c_1X_1 + c_2X_2 + \cdots + c_nX_n \tag{6}$$

s.t.

Technological Constraints:

$$a_{i1}X_1 + a_{i2}X_2 + \ldots + a_{in}X_n \begin{pmatrix} \leq \\ = \\ \geq \end{pmatrix} b_i, \quad i = 1, \ldots, m \tag{7}$$

Sign Restrictions:

$$(X_j \geq 0) \text{or} (X_j \leq 0) \text{or} (X_j \text{ urs}), \qquad j = 1, \ldots, n \tag{8}$$

where "urs" implies unrestricted in sign.

The formulation of Equations 6 to 8 has the general structure of a mathematical programming problem, presented in the introduction of this section, but it is further characterized by the fact that the functions involved in the problem objective and the left-hand-side of the technological constraints are linear. It is the assumptions implied by linearity that to a large extent determine the applicability of the above model in real-world applications.

To provide a better feeling of the linearity concept, let us assume that the different decision variables $X_1, \ldots, X_n$ correspond to various activities from which any solution will be eventually synthesized, and the values assigned to the variables by any given solution indicate the activity level in the considered plan(s). For instance, in the above example, the two activities are the production of items P1 and P2 , while the activity levels correspond to the daily production volume. Furthermore, let us assume that each technological constraint of Equation 7 imposes some restriction on the consumption of a particular resource. Referring back to the prototype example, the two problem resources are the daily production capacity of the two workstations W1 and W2. Under this interpretation, the linearity property implies that:

Additivity assumption: the total consumption of each resource, as well as the overall objective value are the aggregates of the resource consumptions and the contributions to the problem objective, resulting by carrying out each activity independently, and

Proportionality assumption: these consumptions and contributions for each activity are proportional to the actual activity level. It is interesting to notice how the above statement reflects to the logic that was applied when we derived the technological constraints of the prototype example: (i) Our assumption that the processing of each unit of product at every station requires a constant amount of time establishes the proportionality property for our model. (ii) The assumption that the total processing time required at every station to meet the production levels of both products is the aggregate of the processing times required for each product if the corresponding activity took place independently, implies that our system has an additive behavior. It is also interesting to see how the linearity assumption restricts the modeling capabilities of the LP framework: As an example, in the LP paradigm, we cannot immediately model effects like economies of scale in the problem cost structure, and/or situations in which resource consumption by one activity depends on the corresponding consumption by another complementary activity. In some cases, one can approach these more complicated problems by applying some linearization scheme. The resulting approximations for many of these cases have been reported to be quite satisfactory.

Another approximating element in many real-life LP applications results from the so called divisibility assumption. This assumption refers to the fact that for LP theory and algorithms to work, the problem variables must be real. However, in many LP formulations, meaningful values for the levels of the activities involved can be only integer. This is, for instance, the case with the production of items P_1 and P_2 in our prototype example. Introducing integrality requirements for some of the variables in an LP formulation turns the problem to one belonging in the class of (Mixed) Integer Programming (MIP). The complexity of a MIP problem is much higher than that of LP's. Actually, the general IP formulation has be shown to belong to the notorious class of NP-complete problems. (This is a class of problems that have been "formally" shown to be extremely ``hard'' computationally). Given the increased difficulty of solving IP problems, sometimes in practice, near optimal solutions are obtained by solving the LP formulation resulting by relaxing the integrality requirements — known as the LP relaxation of the corresponding IP — and (judiciously) rounding off the fractional values for the integral variables in the optimal solution. Such an approach can be more easily justified in cases where the typical values for the integral variables are in the order of tens or above, since the errors introduced by the rounding-off are rather small, in a relative sense.

We conclude our discussion on the general LP formulation, by formally defining the solution search space and optimality. Specifically, we shall define as the feasible region of the LP of Equations 6 to 8, the entire set of vectors $< X_1, X_2, \ldots, X_n >^T$ that satisfy the technological constraints of Eq. 7 and the sign restrictions of Eq. 8. An optimal solution to the problem is any feasible vector that further satisfies the optimality requirement expressed by Eq. 6. In the next section, we provide a geometric characterization of the feasible region and the optimality condition, for the special case of LP's having only two decision variables.

2.5.4 Optimal Solution by Graphical Method

This section develops a solution approach for LP problems, which is based on a geometrical representation of the feasible region and the objective function. In particular, the space to be considered is the n-dimensional space with each dimension defined by one of the LP variables Xj . The objective function will be described in this n-dim space by its contour plots, *i.e.*, the sets of points that correspond to the same objective value. To the extent that the proposed approach requires the visualization of the underlying geometry, it is applicable only for LP's with upto three variables. Actually, to facilitate the visualization of the concepts involved, in this section we shall restrict ourselves to the two-dimensional case, *i.e.*, to LP's with two decision variables. In the next section, we shall generalize the geometry introduced here for the 2-var case, to the case of LP's with *n* decision variables, providing more analytic (algebraic) characterizations of these concepts and properties.

2.5.4.1 Graphical Solution of the Prototype Example 1: 2-var LP with a unique optimal solution

The sliding motion described suggests a way for identifying the optimal values for, let's say, a max LP problem. The underlying idea is to keep "sliding" the is profit line $c_1X_1 + c_2X_2 = \alpha_0$. in the direction of increasing α's, until we cross the boundary of the LP feasible region. The implementation of this idea on the prototype LP of equation 5 is depicted in Figure 2.1.

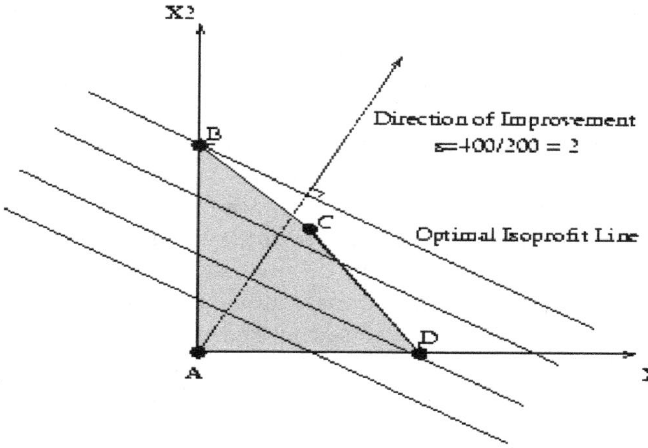

Figure 2.1: Graphical Solution of the Prototype Example LP.

From Figure 2.1, it follows that the optimal daily production levels for the prototype LP are given by the coordinates of the point corresponding to the intersection of line $\frac{1}{30}X_1 + \frac{1}{30}X_2 = 0$ with the X_2-axis, *i.e.*, $X_1^{opt} = 0$; $X_2^{opt} = 50$. The maximal daily profit is $f(X_1^{opt}, X_2^{opt}) = 200.0 + 400.50 = 20,000 (\$)$.

Notice that the optimal point is one of the ``corner'' points of the feasible region depicted in Figure. Can you argue that for the geometry of the feasible region for 2-var LP's described above, if there is a bounded optimal solution, then there will be one which corresponds to one of the corner points?

2.5.4.2 Graphical Solution of the Prototype Example 2: 2-var LP's with many optimal solutions

Consider our prototype example with the unit profit of item P_1 being \$600 instead of \$200. Under this modification, the problem is profit lines become:

$$600X_1 + 400X_2 = \alpha \Leftrightarrow X_2 = -\frac{3}{2}X_1 + \frac{\alpha}{400}$$

and they are parallel to the line corresponding to the first problem constraint:

$$\frac{1}{40}X_1 + \frac{1}{60}X_2 = 1 \Leftrightarrow X_2 = -\frac{3}{2}X_1 + 60.$$

Therefore, if we try to apply the optimizing technique of the previous paragraph in this case, we get the situation depicted below, *i.e.*, every point in the line segment CD is an optimal point, providing the optimal objective value of \$24,000.

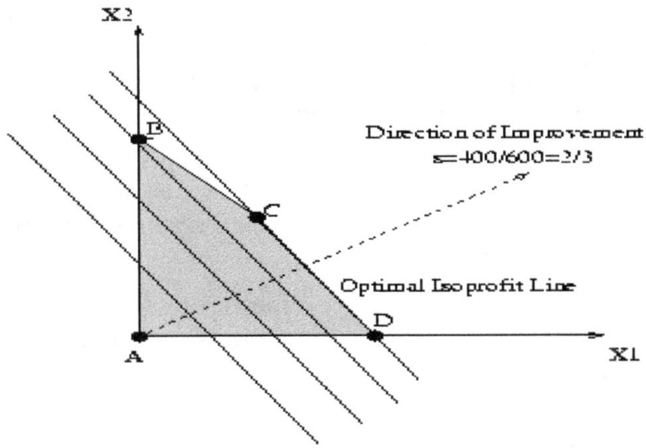

Figure 2.2: An LP with many Optimal Solutions.

It is worth-noticing that even in this case of many optimal solutions, we have two of them corresponding to ``corner'' points of the feasible region, namely points C and D.

2.5.5 Game Concept

Game theory attempts to mathematically capture behavior in strategic situations, in which an individual's success in making choices depends on the choices of others. While initially developed to analyze competitions in which one individual does better at another's expense (zero sum games), it has been expanded to treat a wide class of interactions, which are classified according to several criteria. Today, "game theory is a sort of umbrella or 'unified field' theory for the rational side of social science, where 'social' is interpreted broadly, to include human as well as non-human players (computers, animals, plants)".

2.5.5.1 Representation of Games

The games studied in game theory are well-defined mathematical objects. A game consists of a set of players, a set of moves (or strategies) available to those players, and a specification of payoffs for each combination of strategies. Most cooperative games are presented in the characteristic function form, while the extensive and the normal forms are used to define non cooperative games.

2.5.5.2 Extensive Form

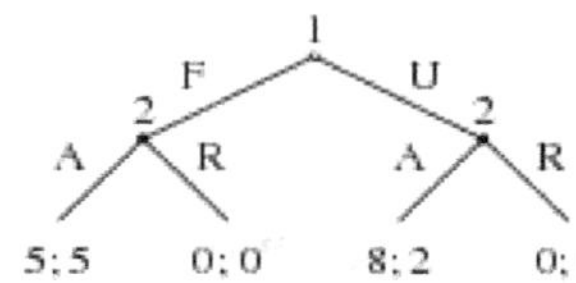

Figure 2.3: Extensive Form Game.

The extensive form can be used to formalize games with some important order. Games here are often presented as trees. Here each vertex (or node) represents a point of choice for a player. The player is specified by a number listed by the vertex. The lines out of the vertex represent a possible action for that player. The payoffs are specified at the bottom of the tree.

The game given in Figure 2.3 here, there are two players. Player 1 moves first and chooses either F or U. Player 2 sees Player 1's move and then chooses A or R. Suppose that Player 1 chooses U and then Player 2 chooses A, then Player 1 gets 8 and Player 2 gets 2.

The extensive form can also capture simultaneous-move games and games with imperfect information. To represent it, either a dotted line connects different vertices to represent them as being part of the same information set (*i.e.*, the players do not know at which point they are), or a closed line is drawn around them.

2.5.5.3 Normal Form Game

The normal (or strategic form) game is usually represented by a matrix which shows the players, strategies, and payoffs. More generally it can be represented by any function that associates a payoff for each player with every possible combination of actions. In the accompanying example there are two players; one chooses the row and the other chooses the column. Each player has two strategies, which are specified by the number of rows and the number of columns. The payoffs are provided in the interior. The first number is the payoff received by the row player (Player 1 in our example); the second is the payoff for the column player (Player 2 in our example). Suppose that Player 1 plays Up and that Player 2 plays Left. Then Player 1 gets a payoff of 4, and Player 2 gets 3.

	Player 2 chooses Left	Player 2 chooses Right
Player 1 chooses Up	4, 3	–1, –1
Player 1 chooses Down	0, 0	3, 4

Figure 2.4: Normal Form or Payoff Matrix of a 2-player, 2-strategy Game.

When a game is presented in normal form, it is presumed that each player acts simultaneously or, at least, without knowing the actions of the other. If players have some information about the choices of other players, the game is usually presented in extensive form.

2.5.5.4 Characteristic Function Form Game

In cooperative games with transferable utility no individual payoffs are given. Instead, the characteristic function determines the payoff of each coalition. The standard assumption is that the empty coalition obtains a payoff of 0.

The origin of this form is to be found in the seminal book of von Neumann and Morgenstern who, when studying coalitional normal form games, assumed that when a coalition C forms, it plays against the complementary coalition (N/C) as if they were playing a 2-player game. The equilibrium payoff of C is characteristic. Now there are different models to derive coalitional values from normal form games, but not all games in characteristic function form can be derived from normal form games.

Formally, a characteristic function form game (also known as a TU-game) is given as a pair (N,v), where N denotes a set of players and $v: 2^N \rightarrow \mathbb{R}$ is a characteristic function. The characteristic function form has been generalised to games without the assumption of transferable utility.

2.5.5.6 Partition Function Form Game

The characteristic function form ignores the possible externalities of coalition formation. In the partition function form the payoff of a coalition depends not only on its members, but also on the way the rest of the players are partitioned.

2.5.6 Game Strategies

The particular behavior or suite of behaviors that a player uses is termed a **strategy**. Strategies can be behaviors that are on some continuum (*e.g.*, how long to wait or display) or they may represent discrete behavior types (*e.g.*, display, fight, or flee). Sometimes the terms **pure strategy** and **mixed strategy** are used.

A **simple or pure strategy** provides a complete definition of how a player will play a game. In particular, it determines the move a player will make for any situation they could face. A player's **strategy set** is the set of pure strategies available to that player. A **pure strategy,** in fact, is a strategy that is not defined in terms of other strategies present in the game.

A **mixed strategy** is an assignment of a probability to each pure strategy. This allows for a player to randomly select a pure strategy. Since probabilities are continuous, there are infinitely many mixed strategies available to a player, even if their strategy set is finite.

Mixed Strategy

Suppose the payoff matrix as given in Figure 2.5 to the right (known as a coordination game). Here one player chooses the row and the other chooses a column. The row player receives the first payoff, the column the second. If row opts to play **A** with probability 1 (*i.e.*, play **A** for sure), then he is said to be playing a pure strategy. If column opts to flip a coin and play **A** if the coin lands heads and **B** if the coin lands tails, then she is said to be playing a mixed strategy, and not a pure strategy.

	A	B
A	1,1	0,0
B	0,0	1,1

Figure 2.5: Pure Coordination Game.

Significance

In his famous paper John Forbes Nash proved that there is an equilibrium for every finite game. One can divide Nash equilibria into two types. Pure strategy Nash equilibria are Nash equilibria where all players are playing pure strategies. Mixed strategy Nash equilibria are equilibria where at least one player is playing a mixed strategy. While Nash proved that every finite game has a Nash equilibrium, not all have pure strategy Nash equilibria. For an example of a game that does not have a Nash equilibrium in pure strategies see Matching pennies. However, many games do have pure strategy Nash equilibria (*e.g.*, the Coordination game, the Prisoner's dilemma, the Stag hunt). Further, games can have both pure strategy and mixed strategy equilibria.

The Nash equilibrium concept is used to analyze the outcome of the strategic interaction of several decision makers. In other words, it is a way of predicting what will happen if several people or several institutions are making decisions at the same time, and if the decision of each one depends on the decisions of the others. The simple insight underlying John Nash's idea is that we cannot predict the result of the choices of multiple decision makers if we analyze those decisions in isolation. Instead, we must ask what each player would do, taking into account the decision-making of the others.

Formal Definition

Let (S, f) be a game with n players, where S_i is the strategy set for player i, $S=S_1 \times S_2 \ldots \times S_n$ is the set of strategy profiles and $f=(f_1(x), \ldots, f_n(x))$ is the payoff function. Let x_{-i} be a strategy profile of all players except for player i. When each player $i \in \{1, \ldots, n\}$ chooses strategy x_i resulting in strategy profile $x = (x_1, \ldots, x_n)$ then player i obtains payoff $f_i(x)$. Note that the payoff depends on the strategy profile chosen, *i.e.*, on the strategy chosen by player i as well as the strategies chosen by all the other players. A strategy profile $x^* \in S$ is a Nash equilibrium (NE) if no unilateral deviation in strategy by any single player is profitable for that player, that is

$$\forall_i, x_i \in S_i, x_i \neq x_i' : f_i(x_i', x_{-i}') \geq f_i(x_i, x_{-i}').$$

A game can have a pure strategy NE or an NE in its mixed extension (that of choosing a pure strategy stochastically with a fixed frequency). Nash proved that if we allow mixed strategies, then every n-player game in which every player can choose from finitely many strategies admits at least one Nash equilibrium.

When the inequality above holds strictly (with > instead of ≥) for all players and all feasible alternative strategies, then the equilibrium is classified as a **strict Nash equilibrium**. If instead, for some player, there is exact equality between x_i' and some other strategy in the set S, then the equilibrium is classified as a **weak Nash equilibrium**.

2.5.6.1 Coordination Game

	Player 2 adopts strategy A	**Player 2 adopts strategy B**
Player 1 adopts strategy A	4, 4	1, 3
Player 2 adopts strategy B	3, 1	3, 3

Figure 2.6: A Sample Coordination Game Showing Relative Payoff for Player 1/Player 2 with each Combination.

The coordination game is a classic (symmetric) two player, two strategy game, with an example payoff matrix shown to the right. The players should thus coordinate, both adopting strategy A, to receive the highest payoff, *i.e.*, 4. If both players chose strategy B though, there is still a Nash equilibrium. Although each player is awarded less than optimal payoff, neither player has incentive to change strategy due to a reduction in the immediate payoff (from 3 to 1). An example of a coordination game is the setting where two technologies are available to two firms with compatible products, and they have to elect a strategy to become the market standard. If both firms agree on the chosen technology, high sales are expected for both firms. If the firms do not agree on the standard technology, few sales result. Both strategies are Nash equilibria of the game.

Driving on a road, and having to choose either to drive on the left or to drive on the right of the road, is also a coordination game. For example, with payoffs 100 meaning no crash and 0 meaning a crash, the coordination game can be defined with the following payoff matrix:

	Drive on the Left	**Drive on the Right**
Drive on the Left	100, 100	0, 0
Drive on the Right	0, 0	100, 100

Figure 2.7: The Driving Game.

In this case there are two pure strategy Nash equilibria, when both choose to either drive on the left or on the right. If we admit mixed strategies (where a pure strategy is chosen at random, subject to some fixed probability), then there are three Nash equilibria for the same case: two we have seen from the pure-strategy form, where the probabilities are (0%,100%) for player one, (0%, 100%) for player two; and (100%, 0%) for player one, (100%, 0%) for player two respectively. We add another where the probabilities for each player is (50%, 50%).

2.5.7 Saddle Point Solution

A saddle point is a point in the domain of a function of two variables which is a stationary point but not a local extremum. At such a point, in general, the surface resembles a saddle that curves up in one direction, and curves down in a different direction (like a mountain pass). In terms of contour lines, a saddle point can be recognized, in general, by a contour that appears to intersect itself. For example, two hills separated by a high pass will show up a saddle point, at the top of the pass, like a figure-eight contour line.

A simple criterion for checking if a given stationary point of a real-valued function $F(x,y)$ of two real variables is a saddle point is to compute the function's Hessian matrix at that point: if the Hessian is indefinite, then that point is a saddle point. For example, the Hessian matrix of the function $z = x^2 - y^2$ at the stationary point (0,0) is the matrix

$$\begin{bmatrix} 2 & 0 \\ 0 & -2 \end{bmatrix}$$

which is indefinite. Therefore, this point is a saddle point. This criterion gives only a sufficient condition. For example, the point (0,0) is a saddle point for the function $z = x^4 - y^4$, but the Hessian matrix of this function at the origin is the null matrix, which is not indefinite.

In the most general terms, a **saddle point** for a smooth function (whose graph is a curve, surface or hyper surface) is a stationary point such that the curve/surface/*etc.* in the neighborhood of that point is not entirely on any side of the tangent space at that point. In one dimension, a saddle point is a point which is both a stationary point and a point of inflection. Since it is a point of inflection, it is not a local extremum.

2.5.7.1 Value and Method of Saddle Point

In mathematics, the steepest descent method or saddle-point approximation is a method used to approximate integrals of the form

$$\int_a^b e^{Mf(x)}dx$$

where f(x) is some twice-differentiable function, M is a large number, and the integral endpoints a and b could possibly be infinite. The technique is also often referred to as Laplace's method, which, in fact, concerns the special case of real-valued functions f admitting a maximum at a real point.

Further, in dynamical systems, a saddle point is a periodic point whose stable and unstable manifolds have a dimension which is not zero. If the dynamic is given by a differentiable map f then a point is hyperbolic if and only if the differential of f^n (where n is the period of the point) has no eigenvalue on the (complex) unit circle when computed at the point.

In a two-player zero sum game defined on a continuous space, the equilibrium point is a saddle point. A saddle point is an element of the matrix which is both the smallest element in its column and the largest element in its row. For a second-order linear autonomous systems, a critical point is a saddle point if the characteristic equation has one positive and one negative real eigenvalue.

2.5.7.2 Laplace's Method

Assume that the function f(x) has a unique global maximum at x_0. Then, the value $f(x_0)$ will be larger than other values f(x). If we multiply this function by a large number M, the gap between $Mf(x_0)$ and Mf(x) will only increase, and then it will grow exponentially for the function $e^{Mf(x)}$.

As such, significant contributions to the integral of this function will come only from points x in a neighborhood of x_0, which can then be estimated.

General Theory of Laplace's Method

To state and prove the method, we need several assumptions. We will assume that x_0 is not an endpoint of the interval of integration, that the values $f(x)$ cannot be very close to $f(x_0)$ unless x is close to x_0, and that $f''(x_0) < 0$.

We can expand f(x) around x_0 by Taylor's theorem,

$$f(x) = f(x_0) + f'(x_0)(x - x_0) + \frac{1}{2}f''(x_0)(x - x_0)^2 + O\left((x - x_0)^3\right).$$

Since f has a global maximum at x_0, and since x_0 is not an endpoint, it is a stationary point, the derivative of f vanishes at x_0. Therefore, the function f(x) may be approximated to quadratic order

$$f(x) \approx f(x_0) - \frac{1}{2}|f''(x_0)|(x - x_0)^2$$

for x close to x_0 (recall that the second derivative is negative at the global maximum $f(x_0)$). The assumptions made ensure the accuracy of the approximation

$$\int_a^b e^{Mf(x)}dx \approx e^{Mf(x_0)}\int_a^b e^{-M|f''(x_0)|(x-x_0)^2/2}dx$$

where the integral is taken in a neighborhood of x_0. This latter integral is a Gaussian integral if the limits of integration go from $-\infty$ to $+\infty$ (which can be assumed so because the exponential decays very fast away from x_0), and thus it can be calculated. We find

$$\int e^{Mf(x)}dx \approx \sqrt{\frac{2\pi}{M|f''(x_0)|}}e^{Mf(x_0)} \text{ as } M \to \infty.$$

2.6. Linear Programming Methods

2.6.1 Standard Form of the Model

Sometimes referred to as the *canonical* form:

$$\text{Maximize } z = \sum_{j=1}^{n} c_j x_j \tag{1}$$

$$\text{subject to } \sum_{j=1}^{n} a_{ij} x_j = b_i, \quad i = 1, \ldots, m \tag{2}$$

$$\text{Minimize} \quad z = x^+ + y^+ - 2x^-$$

1. All RHS parameters $b_i \geq 0$ and $n > m$.
2. Use the n-dimensional vector **x** to represent the decision variables; *i.e.*, **x** $= (x_1, \ldots, x_n)$.
3. Might have simple upper bounds, say, $x_j \leq u_j$.
4. Convert inequalities to equalities in (2).
5. Vector form of constraint: *ai*1*x*1 + • • • + *ainxn* = b_i ; $\mathbf{a}_i\mathbf{x} = b_i$; $\mathbf{Ax} = \mathbf{b}$

Maximize $\{z = \mathbf{cx} : \mathbf{Ax} = \mathbf{b}, \mathbf{x} \geq \mathbf{0}\}$

Definition 1: Let P be a polyhedron in n-dimensional space, written $P \subset Rn$. A vector $\mathbf{x} \in P$ is an *extreme point* of P is we cannot find two vectors $\mathbf{y}, \mathbf{z} \in P$, both different from **x**, and a scalar $\lambda \in [0, 1]$, such that $\mathbf{x} = \mathbf{ly} + (1 - \lambda)\mathbf{z}$.

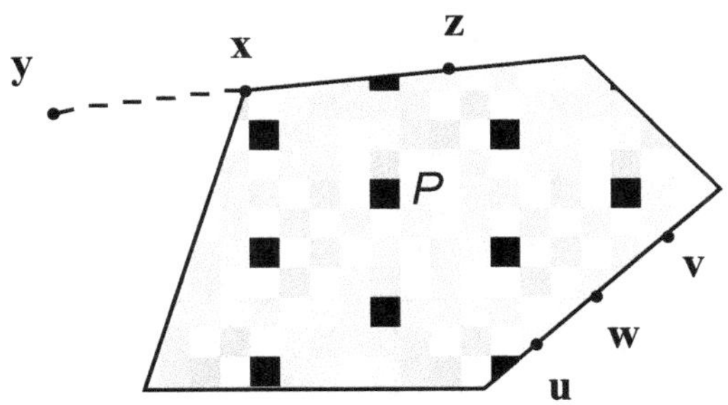

Figure 2.8: Some Points in a Polyhedron.

Definition 2: Let **a** be a nonzero row vector in n-dimensional space, written $\mathbf{a} \in R^n$, and let b be a scalar.

(a) The set $\{\mathbf{x} \in R^n : \mathbf{ax} = b\}$ is called a hyperplane.

(b) The set $\{\mathbf{x} \in R^n : \mathbf{ax} \geq b\}$ is called a halfspace.

Note that a hyperplane is the boundary of a corresponding halfspace. For a polyhedron in n-dimensional space, an extreme point is the intersection of $n - 1$ hyperplanes.

2.6.2 Preparing the Model

- The objective must be to maximize.
- The objective function must be linear in the variables and not contain any constant terms.
- All variables must be restricted to be non-negative.
- Each constraint must be written as a linear equation with the variables on the left of the equality sign and a positive constant on the right.

1. Minimize $\{c_1x_1 + c_2x_2 + \ldots + c_nx_n\} \Leftrightarrow$ Maximize $\{-c_1x_1 - c_2x_2 - \ldots - c_nx_n\}$
2. Constant term in objective function $\Rightarrow$ ignore
3. Nonzero lower bounds on variables: $x_j \geq l_j \Rightarrow x_j = {}_j + l_j$ where ${}_j \geq 0$
4. Unrestricted variables: $x_j = x - x$, where $x \geq 0$ and $x \geq 0$.
5. Constraints not in correct form, $b_i < 0$: multiple by -1

Example

Minimize $z = x + y$	
subject to	$x + y \leq 20$
	$x + y \geq -20$
	$x - y \leq 20$
	$x - y \geq -20$
	x and y unrestricted

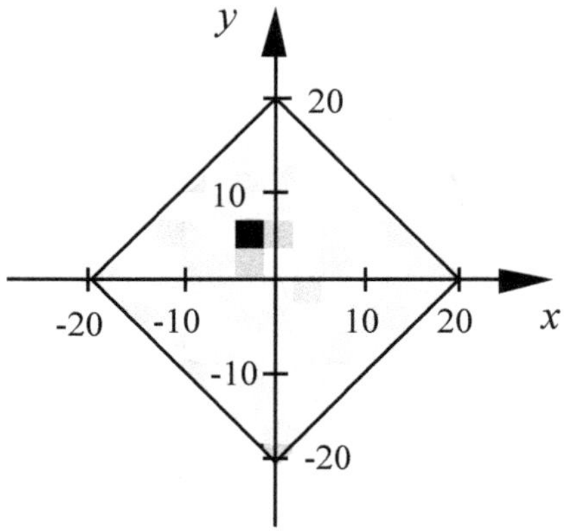

Figure 2.9: Feasible Region for Transformation Example.

Make the substitutions: $x = x+ - x-$ and $y = y+ - x-$, with $x+ \geq 0$, $y+ \geq 0$ and $x- \geq 0$.

At this point, the model is

$$\begin{aligned}
\text{Minimize} \quad & z = x^+ + y^+ - 2x^- \\
\text{subject to} \quad & x^+ + y^+ - 2x^- + s_1 = 20 \\
& x^+ + y^+ - 2x^- - s_2 = -20 \\
& x^+ - y^+ + s_3 = 20 \\
& x^+ - y^+ - s_4 = -20
\end{aligned}$$

$x^+ \geq 0, y^+ \geq 0, x^- \geq 0, s_i \geq 0, i = 1,\ldots,4$

Convert the objective to maximization by multiplying its coefficients by –1.

Multiply both sides of the second and fourth constraint by –1.

$$\begin{aligned}
\text{Maximize} \quad & z = -x^+ - y^+ + 2x^- \\
\text{subject to} \quad & x^+ + y^+ - 2x^- + s_1 = 20 \\
& -x^+ - y^+ + 2x^- + s_2 = 20 \\
& x^+ - y^+ + s_3 = 20 \\
& -x^+ + y^+ + s_4 = 20
\end{aligned}$$

$x^+ \geq 0, y^+ \geq 0, x^- \geq 0, s_i \geq 0, i = 1,\ldots,4$

Solution: $x^- = 10, s_1 = 40, s_3 = 20, s_4 = 20, z = 20 \Rightarrow x = -10, y = -10, z = -20$

Note that this problem has alternate optima. Any point on the line $x + y = -20$ yields the objective value $z = -20$.

2.6.3 Geometric Properties of Linear Programs

Linear Independence

Let us consider a system of m linear equations in m unknowns:

$a11x1 + a12x2 + \bullet\ \bullet\ \bullet + a1mxm = b1$

$a21x1 + a22x2 + \bullet\ \bullet\ \bullet + a2mxm = b2$

$\vdots$

$am1x1 + am2x2 + \bullet\ \bullet\ \bullet + ammxm = bm$

1. Using vector notation, $\mathbf{Ax} = \mathbf{b}$, where $\mathbf{A}$ is an $m \times m$ matrix and $\mathbf{b}$ is an m-dimensional column vector.
2. For this system to have a unique solution, the matrix $\mathbf{A}$ must be *invertible* or *nonsingular*.
3. Must exist another $m \times m$ matrix $\mathbf{D}$ such that $\mathbf{AD} = \mathbf{DA} = \mathbf{I}$, where $\mathbf{I}$ is the $m \times m$ identity
4. Matrix $\mathbf{D}$ is called the *inverse* of $\mathbf{A}$ and is unique. It is denoted by $\mathbf{A}^{-1}$.

Definition 3: Let $\mathbf{A}_1,\ldots,\mathbf{A}_k$ be a collection of k column vectors, each of dimension m. We say that these vectors are *linearly independent* if it is not possible to find k real numbers $a_1, a_2,\ldots,a_k$ not all zero such that $\Sigma\, a_j\mathbf{A}_j = \mathbf{0}$, where $\mathbf{0}$ is the k-dimensional null vector; otherwise, they are called *linearly dependent*.

Theorem 1: Let $\mathbf{A}$ be a square matrix. The following statements are equivalent.

a) The matrix $\mathbf{A}$ is invertible as is its transpose $\mathbf{A}^T$.
b) The determinant of $\mathbf{A}$ is nonzero.
c) The rows and columns of $\mathbf{A}$ are linearly independent.
d) For every vector $\mathbf{b}$, the linear system $\mathbf{Ax} = \mathbf{b}$ has a unique solution.

Assuming that $\mathbf{A}^{-1}$ exists, a popular approach to solving the system $\mathbf{Ax} = \mathbf{b}$ and thus obtaining the unique solution $\mathbf{x} = \mathbf{A}^{-1}\mathbf{b}$ is Gauss-Jordan elimination. The simplex algorithm is based on this approach.

Basic Solutions: Let the matrix $\mathbf{B} = (\mathbf{A}_1,\mathbf{A}_2,\ldots,\mathbf{A}_m)$ be nonsingular. Then we can uniquely solve the equations $\mathbf{Bx}_B = \mathbf{b}$ for the m-dimensional vector $\mathbf{x}_B = (x_1,\ldots,x_m)$. By putting $\mathbf{x} = (\mathbf{x}_B, \mathbf{0})$, that is, by setting the first m-components of $\mathbf{x}$ to those of $\mathbf{x}_B$ and the remaining $n - m$ components to zero, we obtain a solution to $\mathbf{Ax} = \mathbf{b}$. This leads to the following definition.

Definition 4: Given a set of m simultaneous linear equations (2) in n unknowns, let **B** be any nonsingular $m \times m$ matrix made up of columns of **A**. If all the $n - m$ components of **x** not associated with columns of **B** are set equal to zero, the solution to the resulting set of equations is said to be a *basic solution* to (2) with respect to the basis **B**. The components of **x** associated with columns of **B** are called *basic variables*.

Definition 5: A *degenerate basic solution* is said to occur if one or more of the basic variables in a basic solution has value zero.

Definition 6: A vector $\mathbf{x} \in S = \{\mathbf{x} \in Rn : \mathbf{Ax} = \mathbf{b}, \mathbf{x} \geq \mathbf{0}\}$ is said to be *feasible* to the linear programming problem in standard form; a feasible solution that is also basic is said to be a *basic feasible solution* (BFS). If this solution is degenerate, it is called a *degenerate basic feasible solution*.

Foundations: We establish the relationship between optimality and basic feasible solutions in the fundamental theorem of linear programming. The results tell us that when seeking a solution to an LP it is only necessary to consider basic feasible solutions.

Theorem 2: Given a linear program in standard form (1) – (3) where **A** is an $m \times n$ matrix of rank m,

(i) if there is a feasible solution, there is a basic feasible solution;

(ii) if there is an optimal feasible solution, there is an optimal basic feasible solution.

Number of bases: = (4)

Many not feasible

Example: Consider following linear program in standard form where slacks x_3, x_4, x_5 have been added to the 3 constraints.

Maximize $z = 2x_1 + 3x_2$

subject to	$-x_1$	$+ x_2$	$+ x_3$	$= 5$	(C1)
	x_1	$+ 3x_2$	$+ x_4$	$= 35$	(C2)
	x_1	$+ x_5$		$= 20$	(C3)

$x_1 \geq 0, x_2 \geq 0, x_3 \geq 0, x_4 \geq 0, x_5 \geq 0$

Each slack variable can be viewed as the distance from any point in the shadowed polyhedron to the constraint associated with that variable.

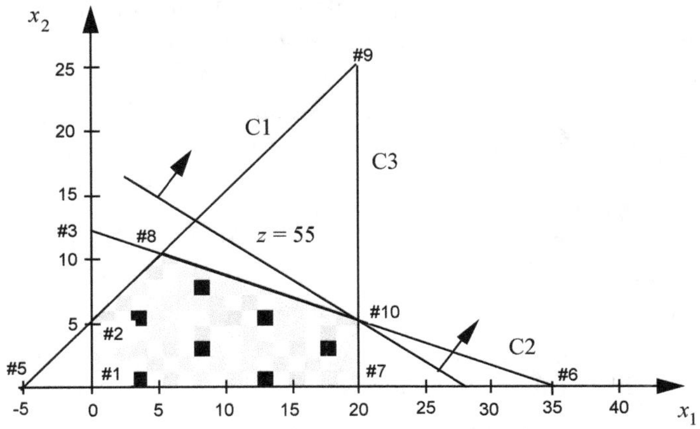

Figure 2.10: Geometry of Example in (x^1,x_2)-Space.

Problem size: n = 5 variables and m = 3 constraints. A basic solution is found by selecting three variables as basic, with the remaining $n - m$ = 2 as nonbasic. For instance, let's try x_1, x_2 and x_3 as basic, and x_4 and x_5 as nonbasic. Setting the nonbasic variables to zero, we obtain the equations

$$\begin{array}{llllll} -x_1 & +x_2 & +x_3 & = & 5 \\ x_1 & +3x_2 & & = & 35 \\ x_1 & & & = & 20 \end{array}$$

which are easily solved to obtain x_1 = 20, x_2 = 5 and x_3 = 20. This is evidently a BFS since all the variables are positive. It is identified on Figure 2.10 as extreme point #10.

For the example, the upper bound on the number of basic solutions is = = 10.

Table 2.1: Basic Solutions to Example

	Basic	*Nonbasic*	*Solution*	*Feasibility*	*Objective*
Point	variables	variables	$(x_1,\ldots,x_5)$	(yes, no)	value
#1	(3,4,5)	(1,2)	(0,0,5,35,20)	Yes	0
#2	(2,4,5)	(1,3)	(0,5,0,20,20)	Yes	15
#3	(2,3,5)	(1,4)	(0,11.7,– 6.7,0,20)	No	—
#4	(2,3,4)	(1,5)	No solution	—	—
#5	(1,4,5)	(2,3)	(– 5,0,0,40,25)	No	—
#6	(1,3,5)	(2,4)	(35,0,40,0,–15)	No	—
#7	(1,3,4)	(2,5)	(20,0,25,15,0)	Yes	40
#8	(1,2,5)	(3,4)	(5,10,0,0,10)	Yes	40
#9	(1,2,4)	(3,5)	(20,25,0,– 60,0)	No	—
#10	(1,2,3)	(4,5)	(20,5,20,0,0)	Yes	55

Theorem 3: For any linear program, there is a unique extreme point in the feasible region **S** = {**x**ÎR*n* : **Ax** = **b**, **x** ≥ **0**} corresponding to a BFS. In addition, there is at least one BFS corresponding to each extreme point in **S**.

Definition 7: For a linear program in standard form with m constraints, two basic feasible solutions are said to be *adjacent* if they have $m - 1$ basic variables in common.

Theorem 4: For a linear program in standard form (1) – (3), let $\mathbf{x}^l$ be a BFS and let $E(l)$ be the index set of all BFSs adjacent to $\mathbf{x}^l$. Define $z^l \equiv z(\mathbf{x}^l)$. If $z^l \geq z^k$ for all $k \in E(l)$ then $\mathbf{x}^l$ is an optimal BFS.

Outline of Simplex Algorithm (max problem)

Step 1: Find an initial BFS to the linear program. As the algorithm progresses, the most recent BFS will be called the incumbent.

Step 2: Determine if the incumbent is optimal. If so, stop; otherwise go to Step 3.

Step 3: Move to an adjacent BFS that has a larger z value and return to Step 2 using the new BFS as the incumbent.

2.6.4 Simplex Tableau

Previous model in slightly modified form → *Simplex Form*:

Maximize z

$$
\begin{array}{llllll}
(E0) & z & -2x_1 & -3x_2 & & = 0 \\
(E1) & & -x_1 & +x_2 & +x_3 & = 5 \\
(E2) & & x_1 & +3x_2 & +x_4 & = 35 \\
(E3) & & x_1 & & +x_5 & = 20
\end{array}
$$

$x_j \geq 0, j = 1,\ldots,5$

Conditions for the Simplex Form

- Every basic variable appears with a nonzero coefficient in one and only one equation and does not appear in (E0).
- In the equation where the basic variable does appear, it has the coefficient 1.
- Each equation includes only one basic variable.
- z appears only in (E0) with the coefficient 1.

The above example happens to be in the simplex form if we identify the basic variables as x_3, x_4 and x_5. When $x_1 = x_2 = 0$, the solution $z = 0$, $x_3 = 5$, $x_4 = 35$, $x_5 = 20$ is easily determined by observation.

New system after algebraic manipulation (elementary row operations):

$$
\begin{array}{llllllll}
(E0') = (E0) + 2(E1') & z & & -5x2 & -2x3 & = & -10 \\
(E1') = -(E1) & & x1 & -x2 & -x3 & = & -5 \\
(E2') = (E2) - (E1') & & 4x2 & +x3 & +x4 & = & 40 \\
(E3') = (E3) - (E1') & & x2 & +x3 & +x5 & = & 25
\end{array}
$$

These equations are entirely equivalent to the original set in that they have the same set of solutions. Note that $(x1, x2, x3, x4, x5) = (0, 0, 5, 35, 20)$ and $z = 0$ is still a solution to the new system. By design, the equations are again in the simplex form where we identify the basic variables as x_1, x_4 and x_5. The basic solution is thus $(x1, x4, x5) = (-5, 40, 25)$, $(x2, x3) = (0, 0)$, $z = -10$.

Which is evidently infeasible because $x_1 < 0$.

2.6.4.1 General Simplex Form

During the simplex computations, we will use $_{ij}$ to denote the current coefficient of x_j in equation i, for $i = 1,\ldots,m$ and $j = 1,\ldots,n$. We use $_j$ in equation (E0) to denote the current value of the jth objective function coefficient. For minimization objective → values are known as *reduced costs* or *relative costs*. They are the unit costs relative to the current set of nonbasic variables in the reduced model in which the basic variables are fixed. The general form of the model is

$$\text{(E0)} \quad z + \sum_{j\in Q} \bar{c}_j x_j = \bar{z}$$

$$\text{(E}i\text{)} \quad x_B(i) + \sum_{j\in Q} \bar{a}_{ij} x_j = \bar{b}_i, \; i = 1,\ldots,M$$

The "bars" on the coefficients $_{ij}$, $_i$, $_j$ indicate that these values are not in general the same as those in the original model (1) – (3).

2.6.4.2 Tableau Description of Equations

To organize the computations, it is common to convert the algebraic model to a tableau, as illustrated in Table for the example problem.

Algebraic Model

(E0)	z	$-2x1$	$-3x2$		$= 0$
(E1)		$-x1$	$+x2$	$+x3$	$= 5$
(E2)		$x1$	$+3x2$	$+x4$	$= 35$
(E3)		$x1$	$+x5$		$= 20$

Table 2.2: Tableau Form of Example Problem, E(0) – E(3)

Row	*Basic*			*Coefficients*				*RHS*
		z	x_1	x_2	x_3	x_4	x_5	
0	z	1	-2	-3	0	0	0	0
1	x_3	0	-1	1	1	0	0	5
2	x_4	0	1	3	0	1	0	35
3	x_5	0	1	0	0	0	1	20

Table 2.3: The Tableau for the Updated Set of Equations (E0′) - (E3′).

Row	*Basic*	Coefficients						*RHS*
		z	x_1	x_2	x_3	x_4	x_5	
0	z	1	0	– 5	– 2	0	0	– 10
1	x_1	0	1	– 1	– 1	0	0	– 5
2	x_4	0	4	1	1	0	0	40
3	x_5	0	0	1	1	0	1	25

The tableau in Table 2.4 gives the general form for a problem with n variables and m constraints. It is assumed that m variables have been singled out as basic.

Table 2.4: General Form of Tableau

Row	*Basic*	z	x_1	x_2	...	x_n	*RHS*
0	z	1	1	2	...	n	
1	$x_{B(1)}$	0	11	12	...	$1n$	1
.	.	.	.	.		.	.
.	.	.	.	.	...	.	.
.	.	.	.	.		.	.
m	$x_{B(m)}$	0	$m1$	$m2$	...	mn	m

2.6.4.3 Interpretation of Coefficients

Written in terms of the basic variables and xk, the equations in the simplex form are

(E0) $z + kxk =$

(Ei) $x\mathrm{B}(i) + ikxk = i,\ i = 1,\ldots,m$

When xk increases by one unit, the value of z must *decrease* by k units (of course, when $k < 0$, z will increase by $-k$ units). Similarly, when xk increases by one unit, the value of $x\mathrm{B}(i)$ must *decrease* by ik. Thus the marginal effects of increasing xk can be described by the following partial derivatives.

$$\frac{\partial z}{\partial x_i} = -\bar{c}_i \text{ and } \frac{\partial x_{B(i)}}{\partial x_i} = -\bar{a}_{i} \text{ for } i = 1,\ldots,m \qquad (6)$$

Referring to the example problem, when the basic variables are x_3, x_4 and x_5, the coefficients in the column for the nonbasic variable x_2 indicate the following effects when x_2 is increased by one unit:

z will increase by 3

x_3 will decrease by 1

x_4 will decrease by 3

x_5 will not change.

2.6.5 Information from the Simplex Form

In general terms, the equations are written as

(E0)z + $\in$ =

(Ei) $xB(i)$ + $\in = i, i = 1,\ldots,m$

For the example B(1) = 3, B(2) = 4, B(3) = 1.

Figure 2.11: Feasible Region.

As an example, we use the linear program given below with its graphical representation shown in above Figure 2.11.

Maximize z =		$3x1$	$+ 2x2$
subject to	$-x1$	$+ 2x2$	≤ 30
	$2x1$	$+ x2$	≤ 40
	$x1$	$+ 2x2$	≥ 10

$x1 \geq 0, \quad x2 \geq 0$

Adding slacks and so forth

$-x1 + 2x2$	$+ x3$	=	30	(E1)
$2x1 + x2$	$+ x4$	=	40	(E2)
$-x1 - 2x2$	$+ x5$	=	-10	(E3)

Table 2.5: Tableau Corresponding to Origin in Figure 2.11

					Coefficients			
Row	*Basic*	*z*	x_1	x_2	x_3	x_4	x_5	*RHS*
0	*z*	1	0	4	0	0	–3	30
1	x_3	0	0	4	1	0	–1	40
2	x_4	0	0	–3	0	1	2	20
3	x_1	0	1	2	0	0	–1	10

Solution from tableau: **x** = (10, 0, 40, 20, 0) with $z = 30$.

From the general expression of the simplex form we read the objective function value and the values of the basic variables as

$z =$, xB(i) = i for $i = 1,\ldots,m$, and $xj = 0$ for $j \in Q$.

Is the basic solution feasible? $i \geq 0$ for $i = 1,\ldots,m$?

The objective function equation for the example is

(E0)$z + 4x2 - 3x5 = 30$

For the current BFS, both $x2$ and $x5$ are zero, so decreasing either causes the solution to become infeasible. We, therefore, restrict attention to increases. When one of the nonbasic variables is held at zero, only z and the other nonbasic variable play a role in (E0).

For $x2$ active, we have $z = 30 - 4x2$.

For $x5$ active, we have $z = 30 + 3x5$

We wish to increase x_5 because we want to maximize the objective function.

Optimality conditions: $j \geq 0$ for all j ÎQ

Condition easily tested by examining equation (E0) in the simplex form or row 0 of the tableau. **How can a nonoptimal feasible solution be improved?** In the general case, all nonbasic variables xj with $j < 0$ are candidates to become basic. How much should the nonbasic variable be changed?

From the example, when x_2 is held at 0, we have

$x3 = 40 + x5$

$x4 = 20 - 2x5$

$x1 = 10 + x5$

As $x5$ –> 10, $x4$ –> 0 which determines how much the nonbasic variable $x5$ can change.

Minimum Ratio Test

$$\theta = \text{minimum}\left\{\frac{\bar{b}_i}{\bar{a}_{ij}} : \bar{a}_{ij} > 0,\ i = 1,\ldots,m\right\}$$

It is possible that no positive value of ij exists for some nonbasic variable xk such that $k < 0$. This implies that xk and z can be increased indefinitely without the solution becoming infeasible. When this occurs, there is no finite optimum so the solution is said to be *unbounded*.

2.6.5.1 Obtaining New Simplex Form

When a nonbasic variable is increased by the value θ, one (or more) of the basic variables goes to zero. → change BFSs → pivot operation → move from one BFS to an adjacent BFS.

For the example, the basic set identified in Table 2.6 is ($x3$,$x4$,$x1$). When $x5$ is selected to increase, we discover that the pivot row is 2. The maximum increase for $x5$ is computed to be 10, causing the basic variable for the second row, $x4$, to go to zero.

Table 2.6: Original and Subsequent Tableaus

I

				Coefficients				
Row	*Basic*	z	x_1	x_2	x_3	x_4	x_5	*RHS*
0	z	1	0	4	0	0	–3	30
1	$x3$	0	0	4	1	0	–1	40
2	$x4$	0	0	–3	0	1	2	20
3	$x1$	0	1	2	0	0	–1	10

Now, divide row 2 by 2 to obtain a new row 2.

II

				Coefficients				
Row	*Basic*	z	x_1	x_2	x_3	x_4	x_5	*RHS*
0	z	1	0	4	0	0	–3	30
1	x3	0	0	4	1	0	–1	40
2	**x5**	0	0	–3/2	0	1/2	**1**	**10**
3	x1	0	1	2	0	0	–1	10

Multiply the new row 2 by +3 and add it to row 0,

Multiply the new row 2 by +1 and add it to row 1,

Multiply the new row 2 by +1 and add it to row 3, to obtain:

III

				Coefficients				
Row	*Basic*	z	x_1	x_2	x_3	x_4	x_5	*RHS*
0	z	1	0	–1/2	0	3/2	0	60
1	$x3$	0	0	5/2	1	–1/2	0	50
2	$x5$	0	0	–3/2	0	1/2	1	10
3	$x1$	0	1	½	0	–1/2	0	20

The third tableau in Table 2.6 is in the simplex form for the basic variables ($x3$, $x5$, $x1$). The BFS is easily seen to be $\mathbf{x} = (20, 0, 50, 0, 10)$ with $z = 60$.

As expected, the objective value has improved by $-5 \times \theta = -3 \times 10 = 30$ to a value of 60. We observe from row 0 that the solution is still not optimal.

Pivot operations: Let the variable entering the basis be *xk* and let the pivot row be *r*. Now,

i) divide row *r* by the pivot element *rk*;
ii) for each row *i*, ($i = 0,\ldots,m$, $i \neq r$), multiply the new row *r* by –*ik* and add it to row *i*.

2.6.5.2 Primal Simplex Algorithm

Step 1 (*Initialization*) *Express the linear program in the equation form*. Choose a subset of the variables that yields a BFS. Transform the equations into the simplex form. Create the first tableau using the coefficients from these equations. If row 0 has all non-negative coefficients, stop with the optimal solution.

Step 2 (*Iteration k*)

a) *Select a variable to enter the basis.*

Find the nonbasic variable that has the most negative reduced cost in row 0 (we are using the steepest ascent rule here). Let that variable be the entering variable and call the corresponding column of the tableau the pivot column. Let the index of the entering variable be *s*.

b) *Find the variable that must leave the basis.*

Compute the ratios between the right-hand-side *i* and the positive elements of the pivot column *is* > 0. Find the row with the minimum ratio and call it the pivot row. Let the index of the pivot row be *r*.

c) *Change the basis.*

Let x_s become the basic variable for row *r* and call $\bar{a}_{rs}$ the pivot element.

Let $\bar{a}_i$ be the vector of coefficients for the *i*th row in the current tableau with right-hand-side $\bar{b}_i$.

Let $\bar{a}_i^{new}$ be the vector of coefficients for the *i*th row in the new tableau with right-hand-side $\bar{b}_i^{new}$.

Create a new tableau by performing the following operations.

For row *r* (divide pivot row by pivot element)

$$\bar{a}_r^{new} = \bar{a}_r / \bar{a}_{rs} \text{ and } \bar{b}_r^{new} = \bar{b}_r / \bar{a}_{rs}$$

For row $i = 0,1,\ldots,m$ and $i \neq r$ (multiply the new pivot row by the negative of the *i*th coefficient in the pivot column and add the result to row *i*)

$$\bar{a}_i^{new} = \left(-\bar{a}_{is} \times \bar{a}_r^{new}\right) + \bar{a}_i \text{ and } \bar{b}_i^{new} = \left(-\bar{a}_{is} \times \bar{b}_r\right) + \bar{b}_i$$

Step 3 (*Optimality test*) If all reduced costs in row 0 are non-negative, stop with the optimal solution. Otherwise return to Step 2.

Example

We will use as an example the problem considered in Section 2.6.4. The tableau in Table 2.6 is repeated in table below with slacks *x*3, *x*4 and *x*5.

Table 2.7: Initial Tableau for Simplex Example

		Coefficients						
Row	*Basic*	*z*	x_1	x_2	x_3	x_4	x_5	*RHS*
0	*z*	1	– 2	– 3	0	0	0	0
1	*x*3	0	– 1	1	1	0	0	5
2	*x*4	0	1	–3	0	1	0	35
3	*x*5	0	1	0	0	0	1	20

Step 1 (*Initialization*)

Note that the initial tableau is in the simplex form. Choosing the slacks x_3, x_4, x_5 as the basic variables gives a BFS. This is a direct result of the original model having all (≤) constraints and positive constants on the right-hand side. In cases such as this, the slack variables will always define a BFS and provide a convenient starting basis for the simplex method. Thus the initialization step is accomplished, for example, from row 0 we see that the solution is not optimal.

Iteration 1

Step 2 (*Iterate - update tableau*)

a) Using the *most negative rule*, we choose x_2 as the entering variable (*s* = 2).

b) We compute the ratios *i* / *is* for positive *is*. For convenience, we show them in a column to the right of the tableau as in table below shown. The index of the pivot row is *r* = 1.

Table 2.8: Tableau with Ratios Computed

		Coefficients							
Row	Basic	*z*	x_1	x_2	x_3	x_4	x_5	RHS	Ratio
0	*z*	1	-2	-3	0	0	0	0	---
1	x_3	0	-1	1	1	0	0	5	5
2	x_4	0	1	3	0	1	0	35	11.67
3	x_5	0	1	0	0	0	1	20	---

c) Using the prescribed linear operations the new tableau appears in next table. The value of the pivot element 12 is 1.

Step 3 (*Optimality test*) One of the reduced costs is negative so we return to Step 2.

Table 2.9: Tableau after Iteration 1

Row	Basic	Coefficients						RHS	Ratio
		z	x_1	x_2	x_3	x_4	x_5		
0	z	1	-5	0	3	0	0	15	---
1	x_2	0	-1	1	1	0	0	5	---
2	x_4	0	4	0	-3	1	0	20	5
3	x_5	0	1	0	0	0	1	20	20

To economize on tables, we show both the results of the current iteration and the selection of the entering and leaving variables for the next iteration in the same tableau. For example, Table 2.9 shows the results of iteration 1 and the selections for iteration 2.

Iteration 2

Step 2 (*Iterate - update tableau*)

a) Only $x1$ has a negative coefficient in row 0 so it is selected to enter the basis ($s = 1$).

b) From the ratios we discover the pivot row is $r = 2$ and the pivot element 21 = 4.

c) Using the prescribed linear operations the new tableau appears in table below.

Table 2.10: Tableau after Iteration 2

Row	Basic	Coefficients						RHS	Ratio
		z	x_1	x_2	x_3	x_4	x_5		
0	z	1	0	0	-0.75	1.25	0	40	---
1	x_2	0	0	1	0.25	0.25	0	10	40
2	x_1	0	1	0	-0.75	0.25	0	5	---
3	x_5	0	0	0	0.75	-0.25	1	15	20

Step 3 (*Optimality test*) One of the coefficients in row 0 is still negative so we return to Step 2.

Iteration 3

Step 2 (*Iterate - update tableau*)

a) Now $x3$ has a negative coefficient in row 0 so we select it to enter the basis ($s = 3$).

b) From the ratio test we discover the pivot row is $r = 3$ and the pivot element, 33 = 0.75.

c) Using the prescribed linear operations the new tableau is given in table below.

Step 3 (*Optimality test*) All reduced costs are non-negative so the current solution is optimal.

Table 2.11: Final Tableau

Row	Basic	Coefficients z	x_1	x_2	x_3	x_4	x_5	RHS	Ratio
0	z	1	0	0	0	1	1	55	---
1	x_2	0	0	1	0	0.333	-0.33	5	---
2	x_1	0	1	0	0	0	1	20	---
3	x_3	0	0	0	1	-0.33	1.333	20	---

2.6.6 Special Situations

Tie for the Entering Variable: Arbitrarily select one (the first).

Tie for the Leaving Variable: this will lead to a degenerate BFS.

Definition 8: A basic solution **x** ∈ R*n* is said to be *degenerate* if the number of structural constraints and non-negativity conditions active at **x** is greater than *n*.

In two dimensions, a degenerate basic solution occurs at the intersection of three or more lines; in three dimensions, a degenerate solution is at the intersection of four or more hyperplanes.

Table 2.12: Initial Tableau for Degeneracy Example

Row	Basic	Coefficients z	x_1	x_2	x_3	x_4	x_5	RHS	Ratio
0	z	1	-3	-4	0	0	0	0	---
1	x_3	0	-1	1	1	0	0	5	5
2	x_4	0	1	3	0	1	0	15	5
3	x_5	0	1	0	0	0	1	20	---

After pivoting to obtain the tableau as shown in Table 2.13 at iteration 1, we see that a degenerate solution has been realized due to the tie in the ratio test.

(*x*1, *x*2, *x*3, *x*4, *x*5) = (0, 5, 0, 0, 20)

Now we must select *x*1 to enter the basis. The ratio is minimum in the degenerate row, thus row 2 is the pivot row and *x*4 leaves the basis.

Table 2.13: Iteration 1

Row	Basic	Coefficients z	x_1	x_2	x_3	x_4	x_5	RHS	Ratio
0	z	1	-7	0	4	0	0	20	---
1	x_2	0	-1	1	1	0	0	5	---
2	x_4	0	4	0	-3	1	0	0	0
3	x_5	0	1	0	0	0	1	20	20

Pivoting we obtain the tableau in below table at iteration 2.

Table 2.14: Iteration 2

Row	Basic	Coefficients						RHS	Ratio
		z	x_1	x_2	x_3	x_4	x_5		
0	z	1	0	0	-1.25	1.75	0	20	---
1	x_2	0	0	1	0.25	0.25	0	5	20
2	x_1	0	1	0	-0.75	0.25	0	0	---
3	x_5	0	0	0	0.75	-0.25	1	20	26.67

The solution is the same but the basis has changed. The entering variable is *x*3. The next solution, in Table 2.15, is not degenerate because the coefficient of *x*3 in row 2, the degenerate row, is negative (–0.75), and there is no tie in the minimum ratio.

Table 2.15: Optimal Solution

Row	Basic	Coefficients						RHS
		z	x_1	x_2	x_3	x_4	x_5	
0	z	1	0	5	0	3	0	45
1	x_3	0	0	4	1	1	0	20
2	x_1	0	1	3	0	1	0	15
3	x_5	0	0	-3	0	-1	1	5

When degeneracy occurs, the possibility arises that the simplex algorithm will return to a previously encountered basis. Nevertheless, cycling can be eliminated by modifying the rules used to select the entering and leaving variables. The rules are based on the indices of the problem variables, *xj* (*j* = 1,…,*n*) and are due to Bland.

Rule for selecting the entering variable: From all the variables having negative coefficients in row 0, select the one with the smallest index to enter the basis.

Rule for selecting the leaving variable: Use the standard ratio test to select the leaving variable, but if there is a tie in the ratio test, select from the tied rows the variable having the smallest index.

Alternative optima: If the optimal tableau shows one or more nonbasic variables with zero coefficients in row 0, there are alternative optimal solutions.

If **x**1,…,**x***s* are optimal BFSs, then so is where

= , = 1, *lk* ≥ 0, *k* = 1,…,*s*

Unbounded solution:

Maximize	*z* =	6*x*1	+ 3*x*2		
subject to			*x*2	≤	5
	–4*x*1	+ 3*x*2	≤	15	
	2*x*1	– 3*x*2	≥	2	

*x*1 ≥ 0, *x*2 ≥ 0

After several iterations, the tableau in Table 2.16 is obtained. Variable $x5$ is slated to enter the basis but it is not possible to perform the ratio test.

Table 2.16. Example of unboundedness

		Coefficients							
Row	Basic	z	x_1	x_2	x_3	x_4	x_5	RHS	Ratio
0	z	1	0	0	12	0	-3	66	---
1	x_2	0	0	1	1	0	0	5	---
2	x_4	0	0	0	3	1	-2	34	---
3	x_1	0	1	0	1.5	0	-0.5	8.5	---

An unbounded solution is probably the result of a poorly specified model.

2.7 The Initial BFS

Add artificial variables: For constraint i the variables will be called ai.

$$\text{Maximize } z = -7x1 + 3x2$$

$$\text{subject to } \quad 4x1 + 8x2 = 30$$

$$-2x1 + x2 \geq 3$$

$$x1 \geq 0,\ x2 \geq 0$$

The equality form of the constraints with slacks and artificial variables added is

$$4x1 + 8x2 + a1 = 30$$

$$-2x1 + x2 - x3 + a2 = 3$$

$$x1 \geq 0,\ x2 \geq 0,\ x3 \geq 0,\ a1 \geq 0,\ a2 \geq 0$$

We now have a BFS composed of artificial variables.

2.7.1 Algorithm for Two-Phase Method

Initial step: Put the problem into equality form adding slack and artificial variables has necessary. Let F be the set of constraints that have artificial variables.

Phase 1: Solve

$$\text{Minimize } w = \left\{ \sum_{i \in F} a_i : \sum_{j=1}^{n} a_{ij} x_j + a_i = b_i, i = 1, \ldots, m; a_i \geq 0, x_j \geq 0, \forall i, j \right\}$$

where ai is present in constraint i only if $i \in F$. If some artificial variables remain in the basis at a zero level, the constraints for which they are basic are redundant and can be deleted.

Phase 2: Delete the artificial variables from the problem and revert to the original objective function

$$\text{Maximize} \quad z = \sum_{j=1}^{n} c_j x_j$$

Starting from the BFS obtained at the termination of phase 1, use the simplex method to solve the original problem to optimality.

Example

Phase 1:	Maximize	$w =$				$-a1$	$-a2$	
Phase 2:	Maximize	$z =$	$-7x1$	$+3x2$				
subject to		$4x1$	$+8x2$		$+a1$			$= 30$
	$-2x1$	$+x2$	$-x3$		$+a2$	$= 3$		

$x1 \geq 0,\ x2 \geq 0,\ x3 \geq 0,\ a1 \geq 0,\ a2 \geq 0$

In constructing the tableau in table below, we introduce row 0′ for the phase 1 objective. Row 0′ is not in the simplex form because the coefficients of the basic variables $a1$ and $a2$ are not zero. To price out these coefficients we subtract row 1 and row 2 from row 0′ with the result shown in row 0 which is now used for the Phase 1 computations.

Table 2.17: Initial Tableau for Phase 1

Row	Basic	Coefficients w	x_1	x_2	x_3	α_1	α_2	RHS	Ratio
0'	w	1	0	0	0	1	1	0	
0	w	1	-2	-9	1	0	0	-33	---
1	α_1	0	4	8	0	1	0	30	3.75
2	α_2	0	-2	1	-1	0	1	3	3

Table 2.18: Iteration 1

Row	Basic	Coefficients w	x_1	x_2	x_3	α_1	α_2	RHS	Ratio
0	w	1	-20	0	-8	0	9	-6	---
1	α_1	0	20	0	8	1	-8	6	0.3
2	x_2	0	-2	1	-1	0	1	3	---

Table 2.19: Optimal Solution for Phase 1

Row	Basic	Coefficients w	x_1	x_2	x_3	α_1	α_2	RHS	Ratio
0	w	1	0	0	0	1	1	0	---
1	x_1	0	1	0	0.4	0.05	-0.4	0.3	---
2	x_2	0	0	1	-0.2	0.1	0.2	3.6	---

Phase 2 is initiated with the basic solution (*x*1, *x*2, *x*3) = (0.3, 3.6, 0) found in phase 1. As shown in table below, we put the phase 2 objective function in row 0′. Again we find that the tableau is not in the simplex form so we price out the nonzero coefficients in the columns for the basic variables. The result is labeled row 0.

Table 2.20: Initial Solution for Phase 2

			Coefficients				
Row	*Basic*	*z*	*x1*	*x2*	*x3*	*RHS*	*Ratio*
0′	*z*	1	7	–3	0	0	
0	*z*	1	0	0	–3.4	8.7	--
1	*x1*	0	1	0	0.4	0.3	0.75
2	*x2*	0	0	1	–0.2	3.6	--

Table 2.21 Optimum Solution for Phase 2

			Coefficients				
Row	Basic	z	x_1	x_2	x_3	RHS	Ratio
0	z	1	8.5	0	0	11.25	---
1	x_3	0	2.5	0	1	0.75	---
2	x_2	0	0.5	1	0	3.75	---

2.8 Dual Simplex Algorithm

The dual simplex algorithm is most suited for problems where an initial dual feasible solution available. It is particularly useful for reoptimizing a problem after a constraint has been added or some parameters have been changed so that the previously optimal basis is no longer feasible.

Step 1 (Initialization)

Start with a dual feasible basis and let $k = 1$. Create a tableau for this basis in the simplex form. If the right-hand side entries are all non-negative, the solution is primal feasible, so stop with the optimal solution.

Step 2 (Iteration k)

a) *Select the leaving variable.* Find a row, call it r, with a negative right-hand-side constant; *i.e.*, $r < 0$. Let row r be the pivot row and let the leaving variable be xB(r). A common rule for choosing r is to select the most negative RHS value; *i.e.*,

   $r = \min\{i : i = 1,\dots,m\}$.

b) *Determine the entering variable.* For each negative coefficient in the pivot row, compute the negative of the ratio between the reduced cost in row 0 and the structural coefficient in row r. If there is no negative coefficient, $rj < 0$, stop; there is no feasible solution.

Let the column with the minimum ratio, designated by the index *s*, be the pivot column; let *xs* is the entering variable. The pivot column is determined by the following ratio test.

$$\frac{-\bar{c}_s}{\bar{a}_{rs}} = \min\left\{\frac{-\bar{c}_j}{\bar{a}_{rj}} : \bar{a}_{rj} < 0, j = 1, \ldots, n\right\}$$

c) *Change the basis*. Replace *x*B(*r*) by *xs* in the basis. Create a new tableau by performing the following operations (these are the same as for the primal simplex algorithm).

Let be the vector of the *i*th row of the current tableau, and let be the *i*th row in the new tableau. Let be the RHS for row *i* in the current tableau, and let be the RHS of the new tableau. Let be the element in the *i*th row of the pivot column *s*.

The pivot row in the new tableau is

$$\bar{a}_r^{new} = \bar{a}_r / \bar{a}_{rs} \text{ and } \bar{b}_r^{new} = \bar{b}_r / \bar{a}_{rs}.$$

The other rows in the new tableau are

$$\bar{a}_i^{new} = \left(-\bar{a}_{is} \times \bar{a}_r^{new}\right) + \bar{a}_i \text{ and}$$

$$\bar{b}_i^{new} = \left(-\bar{a}_{is} \times \bar{b}_r^{new}\right) + \bar{b}_i \text{ for } i = 0, 1, \ldots, m, \ i \neq r$$

(These operations have the effect of pricing out the pivot column. Its replacement will have a single 1 in row *r* and a zero in all other rows as required by the simplex form.)

Step 3 (Feasibility test)

If all entries on the right-hand side are non-negative the solution is primal feasible, so stop with the optimal solution. Otherwise, put *k* ◎ *k* +1 and return to Step 2.

2.9 Simplex Method Using Matrix Notation

Decision variables: **x** = (*x*1,…,*xn*)T

Objective coefficients: **c** = (*c*1,…,*cn*)

Right-hand-side constants: **b** = (*b*1,…,*bm*)T

Structural coefficients:

$$A = \begin{pmatrix} a_{11} & a_{12} & \cdots & a_{1n} \\ a_{21} & a_{22} & \cdots & a_{2n} \\ \vdots & & & \vdots \\ a_{m1} & a_{m2} & \cdots & a_{mn} \end{pmatrix}$$

Making use of this notation, can be rewritten as

Maximize $z =$ **cx**

subject to **Ax = b**

$\mathbf{x} \geq \mathbf{0}$

Example

Maximize $z = 2x1 + 1.25x2 + 3x3$

subject to $2x1 + x2 + 2x3 \leq 7$

$3x1 + x2 \leq 6$

$x2 + 6x3 \leq 9$

$x1 \geq 0, x2 \geq 0, x3 \geq 0$

With $x4$, $x5$ and $x6$ as slacks, the matrices and vectors defining the equality form of the model are:

$\mathbf{x} = (x1, x2, x3, x4, x5, x6)T$

$\mathbf{c} = (2, 1.25, 3, 0, 0, 0)$

$$A = \begin{pmatrix} 2 & 1 & 2 & 1 & 0 & 0 \\ 3 & 1 & 0 & 0 & 1 & 0 \\ 0 & 1 & 6 & 0 & 0 & 1 \end{pmatrix} \text{ and } b = \begin{pmatrix} 7 \\ 6 \\ 9 \end{pmatrix}$$

2.9.1 Computing a Basic Solution

Let **x** = (**x**B, **x**N), where **x**B and **x**N refer to the basic and nonbasic variables, respectively.

Let **A** = (**B**, **N**), where **B** is the $m \times m$ basis matrix and **N** is $m \times (n - m)$.

The equations **Ax** = **b** can thus be written

$$(B, N)\begin{pmatrix} x_B \\ x_N \end{pmatrix} = Bx_B + Nx_N = b.$$

Multiplying through be **B**–1 gives

$$\mathbf{x}_B + \mathbf{B}^{-1}\mathbf{N}\mathbf{x}_N = \mathbf{B}^{-1}\mathbf{b}.$$

Solving this set of equations allows us to express the basic variables in terms of the nonbasic variables.

$$\mathbf{x}_B = \mathbf{B}^{-1}(\mathbf{b} - \mathbf{N}\mathbf{x}_N).$$

The objective function can be written as the sum of two terms, one for the basic variables and the other for the nonbasic variables:

$z = \mathbf{c}_B\mathbf{x}_B + \mathbf{c}_N\mathbf{x}_N = \mathbf{c}_B\mathbf{B}^{-1}(\mathbf{b} - \mathbf{N}\mathbf{x}_N) + \mathbf{c}_N\mathbf{x}_N$ (7)

A basic solution is found by setting the nonbasic variables to zero, **x**N = **0**, so

$\mathbf{x}_B = \mathbf{B}^{-1}\mathbf{b}$ and $z = \mathbf{c}_B\mathbf{B}^{-1}\mathbf{b}$

For convenience and later use in the section on the revised simplex method, we introduce the *m*-dimensional row vector of *dual variables*, π, and define it as

$\pi \equiv \mathbf{c}_B\mathbf{B}^{-1}$

so $z = \pi b$.

For the example, when **x**B = (*x*1,*x*3,*x*5)T, the basic solution is

$$x_B = B^{-1}b = \begin{pmatrix} 0.5 & 0 & -0.166 \\ 0 & 0 & 0.166 \\ -1.5 & 1 & 0.5 \end{pmatrix}\begin{pmatrix} 7 \\ 6 \\ 9 \end{pmatrix} = \begin{pmatrix} 2 \\ 1.5 \\ 0 \end{pmatrix}$$

with objective value

$$z = c_B x_B = (2,3,0)\begin{pmatrix} 2 \\ 1.5 \\ 0 \end{pmatrix} = 8.5$$

and dual solution

$$\pi = (\pi_1, \pi_2, \pi_3) = c_B B^{-1}$$

$$= (2,3,0)\begin{pmatrix} 0.5 & 0 & -0.166 \\ 0 & 0 & 0.166 \\ -1.5 & 1 & 0.5 \end{pmatrix} = (1,0,1/6).$$

2.9.2 Marginal Information Concerning the Objective

From Eq. (7) we see that the objective function value for a given basis can be written as

$$z = c_B B^{-1} b + (c_N - c_B B^{-1} N) x_N$$

For consistency with the material in the sections on the simplex tableau, we will work with the above equation in the following form:

$$z = c_B B^{-1} b + (c_N - c_B B^{-1} N) x_N$$

We want to hold all the nonbasic variables to zero except one, *xk*, and observe the effect. The column for *xk* in **N** is the same as the column for *xk* in **A**; call it **A***k*. The objective value as a function of *xk* alone is

$$z = c_B B^{-1} b - (c_B B^{-1} A_k - c_k)^{\wedge} k.$$

The first term on the right of this expression is a constant for a given basis. The coefficient of *xk* in the second term (without the "–" sign) is the marginal change in the objective function for a unit increase in *xk*. Define for the general nonbasic variable *xk*, the marginal change

$$\bar{c}_1 = c_B B^{-1} A_1 - c_1 - c_1 = \pi A_1 - c_1 \tag{8}$$

where *k* is what we called the reduced cost of *xk*.

A sufficient condition for a given basis to be optimal is that increasing any nonbasic variable will cause the object to decrease or stay the same.

$k \geq 0$ for all $k \in Q$

If this condition is not satisfied, every nonbasic variable with a negative marginal value is a candidate to enter the basis.

For the example, the nonbasic variables are $\mathbf{x}_N = (x_2, x_4, x_6)$ with $\mathbf{c}_N = (1.25, 0, 0)$ and π = (1, 0, 1/6). The reduced cost for each of the nonbasic variables is computed using Eq. (8) as follows:

for *x*2 we have 2 = (1, 0, 1/6) (1, 1, 1)T – 1.25 = –1/12,

for *x*4 we have 4 = (1, 0, 1/6) (1, 0, 0)T – 0 = 1,

for *x*6 we have 6 = (1,0,1/6) (0, 0, 1)T – 0 = 1/6.

2.9.3 Matrix Representation of the Simplex Tableau

Given a basis and basis inverse, the matrix representation of an LP can be written as

$$\begin{pmatrix} 1 & c_B B^{-1} & A - c \\ 0 & B^{-1} A & \end{pmatrix} \begin{pmatrix} z \\ x \end{pmatrix} = \begin{pmatrix} c_B B^{-1} b \\ B^{-1} b \end{pmatrix} \tag{9}$$

which is equivalent to equations (E0), (E*i*), *i* = 1,…,*m*, in Section 3.4. The RHS of (9) is $(z, \bar{\mathbf{b}})$.

2.10 Revised Simplex Method

Not necessary to keep track of the entire tableau, on current basis inverse. Following information required.

- *Primal variables:* xB = **B**–1**b**
- *Dual variables:* π = **c**B**B**–1
- *Marginal cost for xk*: *k* = πA*k* – *ck* (**A***k* is the *k*th column of **A**)
- *Pivot column: k* = **B**–1**A***k*

2.10.1 Statement of the Algorithm

Algorithm presented in parallel with the example shown where starting tableau is given in Table 2.22.

Table 2.22: Tableau for Example

Row	Basic	Coefficients							RHS
		z	x_1	x_2	x_3	x_4	x_5	x_6	
0	z	1	0	-0.083	0	1	0	0.1667	8.5
1	x_1	0	1	0.3333	0	0.5	0	-0.167	2
2	x_3	0	0	0.1667	1	0	0	0.1667	1.5
3	x_5	0	0	0	0	-1.5	1	0.5	0

Let,

xB = (x1, x3, x5)T, **c**B = (2, 3, 0),

B = and **B**–1 = .

Step 1. Compute the basic solution.

For the current basis **B**, compute $\mathbf{B}^{-1}$ and the primal and dual solutions $\mathbf{x}_B$ = **B**–1**b**, $\pi = \mathbf{c}_B$**B**–1.	For the example, using $\mathbf{B}^{-1}$ above, we compute $\mathbf{x}_B = (x_1, x_3, x_5)$T = (2, 3/2, 0)T $\pi = (\pi_1, \pi_2, \pi_3) = (1, 0, 1/6)$.

Step 2. Select the variable to enter the basis.

For each nonbasic variable, compute the reduced cost $j = \pi Aj - cj$ If each of these values in non-negative, stop with the optimal solution and compute $z^* = \mathbf{c}_B\mathbf{x}_B$. Otherwise, select the variable with most negative reduced cost to enter the basis. The entering variable is x_s.	As we have seen, for x_2, 2 = (1, 0, 1/6) (1, 1, 1)^T – 1.25 = –1/12, for x_4, 4 = (1, 0, 1/6) (1,0,0)^T – 0 = 1, and for x_6, 6 = (1,0,1/6) (0, 0, 1)^T – 0 = 1/6. Since the reduced cost for x_2 is negative, the solution can be improved by allowing x_2 to enter the basis (s = 2).

Step 3. Compute the pivot column.

Let column s of the matrix **A** be the vector $\mathbf{A}_s$. Compute the pivot column $s = \mathbf{B}^{-1}\mathbf{A}_s$.	For s = 2, the pivot column is 2 = $\mathbf{B}^{-1}$(1, 1, 1)^T = (1/3, 1/6, 0)^T.

Step 4. Find the variable to leave the basis.

Find the row r for which the minimum ratio is obtained; *i.e.*, $\theta = x_{B(r)}/\bar{a}_{rs}$ $= \min\left\{\frac{xb_{(r)}}{\bar{a}_{rs}} : \bar{a}_{rs} > 0\right\}$ If there are no rows that have *is* > 0, the solution is unbounded. Otherwise, the basic variable $x_B(r)$ leaves the basis; θ is the amount that the variable *xs* is to increase in order drive $x_B(r)$ to zero.	At the current iteration, $\theta = \min\left\{\frac{2}{1/3}, \frac{3/2}{1/6}\right\} = 6,$ with the minimum obtained for row 1. Thus x_1 will leave the basis as x_2 increases from 0 to 6. Note that x_1 is the basic variable for row 1. It is important to keep tract of this correspondence.

Step 5. Change the basis.

Replace $x_B(r)$ by *xs* in set of basic variables and replace $c_B(r)$ by *cs* in $\mathbf{c}_B$. Update the basis inverse (this is equivalent to pivoting in the simplex tableau). Return to Step 1.	For the new basis $\mathbf{x}_B = (x_2, x_3, x_5)$T, $\mathbf{c}_B = (1.25, 3, 0)$ The new basis inverse is $\mathbf{B}^{-1} =$. We now go to Step 1.

The algorithm continues until the optimality condition is met at Step 2 or an unbounded solution is discovered at Step 4. For the example problem, the new BFS is solution is discovered at Step 4. For the example problem, the new BFS is

xB = (*x*2, *x*3, *x*5)T = (6, 1/2, 0)T with *z* = **cBxB** = 9.

which is an improvement over the initial basic BFS whose objective function value is *z* = 8.5, as seen in Table 2.23.

Table 2.23: Tableau for Example

		Coefficients							
Row	Basic	z	x_1	x_2	x_3	x_4	x_5	x_6	RHS
0	z	1	0	-0.083	0	1	0	0.1667	8.5
1	x_1	0	1	0.3333	0	0.5	0	-0.167	2
2	x_3	0	0	0.1667	1	0	0	0.1667	1.5
3	x_5	0	0	0	0	-1.5	1	0.5	0

UNIT 3

Polyhedral Methods

3.1 Introduction

A Polyhedron represents a rational polyhedron. It can be bounded or unbounded, need not be full dimensional or may contain a proper affine subspace. It can be empty or zero dimensional. It is saved as a hash table which contains the vertices, generating rays, and the basis of the linearity space of the Polyhedron as well as the defining affine half-spaces and hyperplanes. The output of a Polyhedron looks like this:

```
i1 : convex Hull(matrix {{0,0,-1,-1},{2,-2,1,-1},{0,0,0,0}},matrix {{1},{0},{0}})

o1 = {ambient dimension => 3 }

dimension of linearity space => 0

dimension of polyhedron => 2

number of facets => 5

number of rays => 1

number of vertices => 4

o1 : Polyhedron
```

This table displays a short summary of the properties of the Polyhedron. Note that the number of rays and vertices are modulo the linearity space. So for example a line in QQ^2 has one vertex and no rays. However, one can not access the above information directly, because this is just a virtual hash table generated for the output. The data defining a Polyhedron is extracted by the functions included in this package. A Polyhedron can be constructed as the convex hull (convexHull) of a set of points and a set of rays or as the intersection (intersection) of a set of affine half-spaces and affine hyperplanes.

For example, consider the square and the square with an emerging ray for the convex hull:

```
i2 : V = matrix {{1,1,-1,-1},{1,-1,1,-1}}

o2 = | 1 1 -1 -1 |
```

```
| 1 -1 1 -1 |
2 4
o2 : Matrix ZZ <--- ZZ
i3 : convexHull V
o3 = {ambient dimension => 2 }
dimension of linearity space => 0
dimension of polyhedron => 2
number of facets => 4
number of rays => 0
number of vertices => 4
o3 : Polyhedron
i4 : R = matrix {{1},{1}}
o4 = | 1 |
| 1 |
2 1
o4 : Matrix ZZ <--- ZZ
i5 : convexHull(V,R)
o5 = {ambient dimension => 2 }
dimension of linearity space => 0
dimension of polyhedron => 2
number of facets => 4
number of rays => 1
number of vertices => 3
o5 : Polyhedron
```

If we take the intersection of the half-spaces defined by the directions of the vertices with 1 we get the cross polytope:

```
i6 : HS = transpose V
o6 = | 1 1 |
| 1 -1 |
| -1 1 |
| -1 -1 |
```

```
4 2
o6 : Matrix ZZ <--- ZZ
i7 : v = R | | R
o7 = | 1 |
| 1 |
| 1 |
| 1 |
4 1
o7 : Matrix ZZ <--- ZZ
i8 : P = intersection(HS,v)
o8 = {ambient dimension => 2 }
dimension of linearity space => 0
dimension of polyhedron => 2
number of facets => 4
number of rays => 0
number of vertices => 4
o8 : Polyhedron
i9 : vertices P
o9 = | -1 1 0 0 |
| 0 0 -1 1 |
2 4
o9 : Matrix QQ <--- QQ
```

This can for example be embedded in 3-space on height 1:

```
i10 : HS = HS | matrix {{0},{0},{0},{0}}
o10 = | 1 1 0 |
| 1 -1 0 |
| -1 1 0 |
| -1 -1 0 |
4 3
o10 : Matrix ZZ <--- ZZ
i11 : HP = matrix {{0,0,1}}
```

```
o11 = | 0 0 1 |
1 3
o11 : Matrix ZZ <--- ZZ
i12 : w = matrix {{1}}
o12 = | 1 |
1 1
o12 : Matrix ZZ <--- ZZ
i13 : P = intersection(HS,v,HP,w)
o13 = {ambient dimension => 3 }
dimension of linearity space => 0
dimension of polyhedron => 2
number of facets => 4
number of rays => 0
number of vertices => 4
o13 : Polyhedron
i14 : vertices P
o14 = | -1 1 0 0 |
| 0 0 -1 1 |
| 1 1 1 1 |
3 4
o14 : Matrix QQ <--- QQ
```

3.2 Working with Polyhedral

We start with a polyhedron in 2-space which is the convex Hull of a given set of points.

```
i1 : V = matrix {{0,2,-2,0},{-1,1,1,1}}
o1 = | 0 2 -2 0 |
| -1 1 1 1 |
2 4
o1 : Matrix ZZ <--- ZZ
i2 : P = convexHull V
o2 = {ambient dimension => 2 }
```

dimension of linearity space => 0

dimension of polyhedron => 2

number of facets => 3

number of rays => 0

number of vertices => 3

o2 : Polyhedron

This gives an overview of the characteristics of the polyhedron. If we want to know more details, we can ask for them.

i3 : vertices P

o3 = | 0 -2 2 |

| -1 1 1 |

2 3

o3 : Matrix QQ <--- QQ

Here we see that the point (0,1) is not a vertex and P is actually a triangle.

i4 : (HS,v) = halfspaces P

o4 = (| -1 -1 |, | 1 |)

| 1 -1 | | 1 |

| 0 1 | | 1 |

o4 : Sequence

This gives the defining affine half-spaces, *i.e.* P is given by all p such that HS*p =< v and that lie in the defining affine hyperplanes. To get the hyperplanes we use:

i5 : hyperplanes P

o5 = (0, 0)

o5 : Sequence

There are none, so the polyhedron is of full dimension. It is also compact, since P has no rays and the linearity space is of dimension zero.

i6 : rays P

o6 = 0

2

o6 : Matrix ZZ <--- 0

i7 : linSpace P

o7 = 0

```
2
o7 : Matrix ZZ <--- 0
```

Furthermore, we can construct the convex hull of a set of points and a set of rays.

```
i8 : R = matrix {{1},{0},{0}}
o8 = | 1 |
| 0 |
| 0 |
3 1
o8 : Matrix ZZ <--- ZZ
i9 : V1 = V | | matrix {{1,1,1,1}}
o9 = | 0 2 -2 0 |
| -1 1 1 1 |
| 1 1 1 1 |
3 4
o9 : Matrix ZZ <--- ZZ
i10 : P1 = convexHull(V1,R)
o10 = {ambient dimension => 3 }
dimension of linearity space => 0
dimension of polyhedron => 2
number of facets => 3
number of rays => 1
number of vertices => 2
o10 : Polyhedron
i11 : vertices P1
o11 = | 0 -2 |
| -1 1 |
| 1 1 |
3 2
o11 : Matrix QQ <--- QQ
```

This polyhedron is not compact anymore and also not of full dimension.

```
i12 : rays P1
o12 = | 1 |
| 0 |
| 0 |
3 1
o12 : Matrix ZZ <--- ZZ
i13 : hyperplanes P1
o13 = (| 0 0 1 |, | 1 |)
o13 : Sequence
```

On the other hand we can construct a polyhedron as the intersection of affine half-spaces and affine hyperplanes.

```
i14 : HS = transpose (V | | matrix {{-1,2,0,1}})
o14 = | 0 -1 -1 |
| 2 1 2 |
| -2 1 0 |
| 0 1 1 |
4 3
o14 : Matrix ZZ <--- ZZ
i15 : v = matrix {{1},{1},{1},{1}}
o15 = | 1 |
| 1 |
| 1 |
| 1 |
4 1
o15 : Matrix ZZ <--- ZZ
i16 : HP = matrix {{1,1,1}}
o16 = | 1 1 1 |
1 3
o16 : Matrix ZZ <--- ZZ
i17 : w = matrix {{3}}
o17 = | 3 |
```

```
1 1
o17 : Matrix ZZ <--- ZZ
i18 : P2 = intersection(HS,v,HP,w)
o18 = {ambient dimension => 3 }
dimension of lineality space => 0
dimension of polyhedron => 2
number of facets => 3
number of rays => 0
number of vertices => 3
o18 : Polyhedron
```

This is a triangle in 3-space with the following vertices.

```
i19 : vertices P2
o19 = | 4 4 2 |
| 9 5 5 |
| -10 -6 -4 |
3 3
o19 : Matrix QQ <--- QQ
```

If we don't intersect with the hyperplane we get a full dimensional polyhedron.

```
i20 : P3 = intersection(HS,v)
o20 = {ambient dimension => 3 }
dimension of linearity space => 1
dimension of polyhedron => 3
number of facets => 3
number of rays => 0
number of vertices => 3
o20 : Polyhedron
i21 : vertices P3
o21 = | 10/9 -2/3 -2/9 |
| -7/9 -1/3 5/9 |
| -2/9 -2/3 4/9 |
3 3
```

o21 : Matrix QQ <--- QQ

i22 : linSpace P3

o22 = | -1 |

| -2 |

| 2 |

3 1

o22 : Matrix ZZ <--- ZZ

Note that the vertices are given modulo the linearity space. Besides constructing polyhedral by hand, there are also some basic polyhedral implemented such as the hypercube, in this case with edge-length four.

i23 : P4 = hypercube(3,2)

o23 = {ambient dimension => 3 }

dimension of linearity space => 0

dimension of polyhedron => 3

number of facets => 6

number of rays => 0

number of vertices => 8

o23 : Polyhedron

i24 : vertices P4

o24 = | -2 2 -2 2 -2 2 -2 2 |

| -2 -2 2 2 -2 -2 2 2 |

| -2 -2 -2 -2 2 2 2 2 |

3 8

o24 : Matrix QQ <--- QQ

Another on is the crossPolytope, in this case with diameter six.

i25 : P5 = crossPolytope(3,3)

o25 = {ambient dimension => 3 }

dimension of lineality space => 0

dimension of polyhedron => 3

number of facets => 8

number of rays => 0

number of vertices => 6

o25 : Polyhedron

i26 : vertices P5

o26 = | -3 3 0 0 0 0 |

| 0 0 -3 3 0 0 |

| 0 0 0 0 -3 3 |

3 6

o26 : Matrix QQ <--- QQ

Furthermore the standard simplex (stdSimplex).

i27 : P6 = stdSimplex 2

o27 = {ambient dimension => 3 }

dimension of linearity space => 0

dimension of polyhedron => 2

number of facets => 3

number of rays => 0

number of vertices => 3

o27 : Polyhedron

i28 : vertices P6

o28 = | 1 0 0 |

| 0 1 0 |

| 0 0 1 |

3 3

o28 : Matrix QQ <--- QQ

Now that we can construct polyhedra, we can turn to the functions that can be applied to polyhedra. First of all, we can apply the convexHull function also to a pair of polyhedra:

i29 : P7 = convexHull(P4,P5)

o29 = {ambient dimension => 3 }

dimension of linearity space => 0

dimension of polyhedron => 3

number of facets => 24

number of rays => 0

number of vertices => 14

o29 : Polyhedron

i30 : vertices P7

o30 = | -3 3 0 0 0 -2 2 -2 2 -2 2 -2 2 0 |

| 0 0 -3 3 0 -2 -2 2 2 -2 -2 2 2 0 |

| 0 0 0 0 -3 -2 -2 -2 -2 2 2 2 2 3 |

3 14

o30 : Matrix QQ <--- QQ

Or we can intersect them by using intersection:

i31 : P8 = intersection(P4,P5)

o31 = {ambient dimension => 3 }

dimension of linearity space => 0

dimension of polyhedron => 3

number of facets => 14

number of rays => 0

number of vertices => 24

o31 : Polyhedron

i32 : vertices P8

o32 = | -1 1 -2 2 -2 2 -1 1 -1 1 0 0 -2 2 0 0 -2 2 0 0 -1 1 0 0 |

| -2 -2 -1 -1 1 1 2 2 0 0 -1 1 0 0 -2 2 0 0 -2 2 0 0 -1 1 |

| 0 0 0 0 0 0 0 0 -2 -2 -2 -2 -1 -1 -1 -1 1 1 1 1 2 2 2 2 |

3 24

o32 : Matrix QQ ← QQ

Furthermore, both functions can be applied to a list containing any number of polyhedral and matrices defining vertices/rays or affine half-spaces/hyperplanes. All of these must be in the same ambient space. For example:

i33 : P9 = convexHull {(V1,R),P2,P6}

o33 = {ambient dimension ⇒ 3 }

dimension of linearity space ⇒ 0

dimension of polyhedron ⇒ 3

number of facets ⇒ 8

number of rays ⇒ 1

number of vertices ⇒ 5

o33 : Polyhedron

i34 : vertices P9

o34 = | 4 4 2 0 -2 |

| 9 5 5 -1 1 |

| -10 -6 -4 1 1 |

3 5

o34 : Matrix QQ ← QQ

Further functions are for example the Minkowski sum (minkowskiSum) of two polyhedra.

i35 : Q = convexHull (-V)

o35 = {ambient dimension ⇒ 2 }

dimension of linearity space ⇒ 0

dimension of polyhedron ⇒ 2

number of facets ⇒ 3

number of rays ⇒ 0

number of vertices ⇒ 3

o35 : Polyhedron

i36 : P10 = P + Q

o36 = {ambient dimension ⇒ 2 }

dimension of linearity space ⇒ 0

dimension of polyhedron ⇒ 2

number of facets ⇒ 6

number of rays ⇒ 0

number of vertices ⇒ 6

o36 : Polyhedron

i37 : vertices P10

o37 = | -4 4 -2 2 -2 2 |

| 0 0 -2 -2 2 2 |

2 6

o37 : Matrix QQ ← QQ

In the other direction, we can also determine all Minkowski summands (see minkSummandCone) of a polyhedron.

```
i38 : (C,L,M) = minkSummandCone P10
o38 = ({ambient dimension ⇒ 6 }, HashTable{0 ⇒
dimension of linearity space ⇒ 0
dimension of the cone ⇒ 4
number of facets ⇒ 6
number of rays ⇒ 5

1 ⇒
2 ⇒
3 ⇒
4 ⇒
------------------------------------------------------------------
{ambient dimension => 2 }}, | 1 0 |)
dimension of linearity space ⇒ 0 | 0 1 |
dimension of polyhedron ⇒ 1 | 1 0 |
number of facets ⇒ 2 | 1 0 |
number of rays ⇒ 0 | 0 1 |
number of vertices ⇒ 2
{ambient dimension ⇒ 2 }
dimension of linearity space ⇒ 0
dimension of polyhedron ⇒ 2
number of facets ⇒ 3
number of rays ⇒ 0
number of vertices ⇒ 3
{ambient dimension ⇒ 2 }
dimension of linearity space ⇒ 0
dimension of polyhedron ⇒ 1
number of facets ⇒ 2
number of rays ⇒ 0
number of vertices ⇒ 2
{ambient dimension ⇒ 2 }
```

dimension of linearity space ⇒ 0

dimension of polyhedron ⇒ 1

number of facets ⇒ 2

number of rays ⇒ 0

number of vertices ⇒ 2

{ambient dimension ⇒ 2 }

dimension of linearity space ⇒ 0

dimension of polyhedron ⇒ 2

number of facets ⇒ 3

number of rays ⇒ 0

number of vertices ⇒ 3

o38 : Sequence

i39 : apply(values L, vertices)

o39 = {| 0 4 |, | 0 4 2 |, | 0 2 |, | 0 2 |, | 0 4 2 |}

| 0 0 | | 0 0 -2 | | 0 2 | | 0 -2 | | 0 0 2 |

o39 : List

Here the polyhedral in the hash table L are all possible Minkowski summands up to scalar multiplication and the columns of M give the minimal decompositions. So the hexagon P10 is not only the sum of two triangles but also the sum of three lines. Furthermore, we can take the direct product of two polyhedral.

i40 : P11 = P * Q

o40 = {ambient dimension ⇒ 4 }

dimension of linearity space ⇒ 0

dimension of polyhedron ⇒ 4

number of facets ⇒ 6

number of rays ⇒ 0

number of vertices ⇒ 9

o40 : Polyhedron

i41 : vertices P11

o41 = | 0 -2 2 0 -2 2 0 -2 2 |

| -1 1 1 -1 1 1 -1 1 1 |

| -2 -2 -2 2 2 2 0 0 0 |

| -1 -1 -1 -1 -1 -1 1 1 1 |

4 9

o41 : Matrix QQ ← QQ

The result is in QQ^4.

i42 : ambDim P11

o42 = 4

To find out more about this polyhedron use for example.

i43 : fVector P11

o43 = {9, 18, 15, 6, 1}

o43 : List

The function fVector gives the number of faces of each dimension, so it has 9 vertices, 18 edges and so on. We can access the faces of a certain co-dimension via:

i44 : L = faces(1,P11)

o44 = {{ambient dimension ⇒ 4 }, {ambient dimension ⇒ 4

dimension of linearity space ⇒ 0 dimension of linearity space ⇒

dimension of polyhedron ⇒ 3 dimension of polyhedron ⇒ 3

number of facets ⇒ 5 number of facets ⇒ 5

number of rays ⇒ 0 number of rays ⇒ 0

number of vertices ⇒ 6 number of vertices ⇒ 6

}, {ambient dimension ⇒ 4 }, {ambient dimension ⇒ 4

0 dimension of linearity space ⇒ 0 dimension of linearity space

dimension of polyhedron => 3 dimension of polyhedron ⇒ 3

number of facets ⇒ 5 number of facets ⇒ 5

number of rays ⇒ 0 number of rays ⇒ 0

number of vertices ⇒ 6 number of vertices ⇒ 6

}, {ambient dimension ⇒ 4 },

⇒ 0 dimension of linearity space ⇒ 0

dimension of polyhedron ⇒ 3

number of facets ⇒ 5

```
number of rays ⇒ 0
number of vertices ⇒ 6
---------------------------------------------------------------------
{ambient dimension ⇒ 4 }}
dimension of linearity space ⇒ 0
dimension of polyhedron ⇒ 3
number of facets ⇒ 5
number of rays ⇒ 0
number of vertices ⇒ 6
o44 : List
i45 : apply(L,vertices)
o45 = {| 0 -2 0 -2 0 -2 |, | 0 2 0 2 0 2 |, | -2 2 -2 2 -2 2 |, |
| -1 1 -1 1 -1 1 | | -1 1 -1 1 -1 1 | | 1 1 1 1 1 1 | |
| -2 -2 2 2 0 0 | | -2 -2 2 2 0 0 | | -2 -2 2 2 0 0 | |
| -1 -1 -1 -1 1 1 | | -1 -1 -1 -1 1 1 | | -1 -1 -1 -1 1 1 | |
---------------------------------------------------------------------
0 -2 2 0 -2 2 |, | 0 -2 2 0 -2 2 |, | 0 -2 2 0 -2 2 |}
-1 1 1 -1 1 1 | | -1 1 1 -1 1 1 | | -1 1 1 -1 1 1 |
-2 -2 -2 2 2 2 | | -2 -2 -2 0 0 0 | | 2 2 2 0 0 0 |
-1 -1 -1 -1 -1 -1 | | -1 -1 -1 1 1 1 | | -1 -1 -1 1 1 1 |
o45 : List
```

We can compute all lattice points of the polyhedron with lattice Points.

```
i46 : L = lattice Points P11
o46 = {| 0 |, | 0 |, | 0 |, | 0 |, | 0 |, | 0 |, | 0 |, | 0 |, | 0
| -1 | | -1 | | -1 | | -1 | | -1 | | -1 | | -1 | | -1 | | -1
| -2 | | -1 | | 0 | | 1 | | 2 | | -1 | | 0 | | 1 | | 0
| -1 | | -1 | | -1 | | -1 | | -1 | | 0 | | 0 | | 0 | | 1
---------------------------------------------------------------------
|, | -1 |, | -1 |, | -1 |, | -1 |, | -1 |, | -1 |, | -1 |, | -1 |, | -1
| | 0 | | 0 | | 0 | | 0 | | 0 | | 0 | | 0 | | 0 | | 0
| | -2 | | -1 | | 0 | | 1 | | 2 | | -1 | | 0 | | 1 | | 0
| | -1 | | -1 | | -1 | | -1 | | -1 | | 0 | | 0 | | 0 | | 1
---------------------------------------------------------------------
```

I, I 0 I, I 0 I, I 0 I, I 0 I, I 0 I, I 0 I,0, I 0 I, I 0 I, I

I I 0 I I 0 I I 0 I I 0 I I 0 I I 0 I I 0 I I 0 I I

I I -2 I I -1 I I 0 I I 1 I I 2 I I -1 I I 1 I I 0 I I

I I -1 I I -1 I I -1 I I -1 I I -1 I I 0 I I 0 I I 1 I I

----

1 I, I 1 I, I 1 I, I 1 I, I 1 I, I 1 I, I 1 I, I 1 I, I 1 I, I -2

0 I I 0 I I 0 I I 0 I I 0 I I 0 I I 0 I I 0 I I 0 I I 1

-2 I I -1 I I 0 I I 1 I I 2 I I -1 I I 0 I I 1 I I 0 I I -2

-1 I I -1 I I -1 I I -1 I I -1 I I 0 I I 0 I I 0 I I 1 I I -1

I, I -2 I, I -2 I, I -2 I, I -2 I, I -1 I, I -1 I, I -1 I, I -1 I, I -1

I I 1 I I 1 I I 1 I I 1 I I 1 I I 1 I I 1 I I 1 I I 1

I I -1 I I 0 I I 1 I I 2 I I -2 I I -1 I I 0 I I 1 I I 2

I I -1 I I -1 I I -1 I I -1 I I -1 I I -1 I I -1 I I -1 I I -1

I, I 0 I, I 0 I, I 0 I, I 0 I, I 0 I, I 1 I, I 1 I, I 1 I, I 1

I I 1 I I 1 I I 1 I I 1 I I 1 I I 1 I I 1 I I 1 I I 1

I I -2 I I -1 I I 0 I I 1 I I 2 I I -2 I I -1 I I 0 I I 1

I I -1 I I -1 I I -1 I I -1 I I -1 I I -1 I I -1 I I -1 I I -1

I, I 1 I, I 2 I, I 2 I, I 2 I, I 2 I, I 2 I, I -2 I, I -1 I, I 0

I I 1 I I 1 I I 1 I I 1 I I 1 I I 1 I I 1 I I 1 I I 1

I I 2 I I -2 I I -1 I I 0 I I 1 I I 2 I I -1 I I -1 I I -1

I I -1 I I -1 I I -1 I I -1 I I -1 I I -1 I I 0 I I 0 I I 0

I, I 1 I, I 2 I, I -2 I, I -1 I, I 0 I, I 1 I, I 2 I, I -2 I, I -1 I,

I I 1 I I 1 I I 1 I I 1 I I 1 I I 1 I I 1 I I 1 I I 1 I

I I -1 I I -1 I I 0 I I 0 I I 0 I I 0 I I 0 I I 1 I I 1 I

I I 0 I I 0 I I 0 I I 0 I I 0 I I 0 I I 0 I I 0 I I 0 I

```
| 0 |, | 1 |, | 2 |, | -2 |, | -1 |, | 0 |, | 1 |, | 2 |}
| 1 | | 1 | | 1 | | 1 | | 1 | | 1 | | 1 | | 1 |
| 1 | | 1 | | 1 | | 0 | | 0 | | 0 | | 0 | | 0 |
| 0 | | 0 | | 0 | | 1 | | 1 | | 1 | | 1 | | 1 |
o46 : List
i47 : #L
o47 = 81
```

Evenmore the tail/recession cone of a polyhedron with tail Cone.

```
i48 : C = tailCone P1
o48 = {ambient dimension => 3 }
dimension of linearity space => 0
dimension of the cone => 1
number of facets => 1
number of rays => 1
o48 : Cone
i49 : rays C
o49 = | 1 |
| 0 |
| 0 |
3 1
o49 : Matrix ZZ <--- ZZ
```

Finally, there is also a function to compute the polar of a polyhedron, *i.e.*, all points in the dual space that are greater than -1 on all points of the polyhedron:

```
i50 : P12 = polar P11
o50 = {ambient dimension => 4 }
dimension of linearity space => 0
dimension of polyhedron => 4
number of facets => 9
number of rays => 0
number of vertices => 6
o50 : Polyhedron
```

i51 : vertices P12

o51 = | 0 -1 1 0 0 0 |

| -1 1 1 0 0 0 |

| 0 0 0 -1 1 0 |

| 0 0 0 -1 -1 1 |

4 6

o51 : Matrix QQ ← QQ

3.3 Polyhedral Techniques in Combinatorial Optimization

3.3.1 Duality

Suppose that we have the following linear program in maximization standard form:

$$\begin{aligned}
\text{maximize} \quad & x_1 + 2x_2 + x_3 + x_4 \\
\text{subject to} \quad & \\
& x_1 + 2x_2 + x_3 \leq 2 \\
& x_2 + x_4 \leq 1 \\
& x_1 + 2x_3 \leq 1 \\
& x_1 \geq 0 \\
& x_2 \geq 0 \\
& x_3 \geq 0
\end{aligned}$$

and that an LP-solver has found for us the solution $x_1 = 1, x_2 := \frac{1}{2}, x_3 : 0, x_4 := \frac{1}{2}$ of cost 2.5. How can we convince ourselves, or another user, that the solution is indeed optimal, without having to trace the steps of the computation of the algorithm?

Observe that if we have two valid inequalities

$a \leq b$ and $c \leq d$

then we can deduce that the inequality

$a + c \leq b + d$

(derived by "summing the left hand sides and the right hand sides" of our original inequalities) is also true. In fact, we can also scale the inequalities by a positive multiplicative factor before adding them up, so for every non-negative values $y_1, y_2 \geq 0$ we also have

$$y_1 a + y_2 c \leq y_1 b + y_2 d$$

Going back to our linear program (1), we see that if we scale the first inequality by $\frac{1}{2}$, add the second inequality, and then add the third inequality scaled by $\frac{1}{2}$, we get that, for every (x_1, x_2, x_3, x_4) that is feasible for (1),

$x_1 + 2x_2 + 1.5x_3 + x_4 \leq 2.5$

And so, for every feasible (x_1, x_2, x_3, x_4), its cost is

$$x_1 + 2x_2 + x_3 + x_4 \leq x_1 + 2x_2 + 1.5x_3 + x_4 \leq 2.5$$

meaning that a solution of cost 2.5 is indeed optimal.

In general, how do we find a good choice of scaling factors for the inequalities, and what kind of upper bounds can we prove to the optimum?

Suppose that we have a maximization linear program in standard form. For every choice of non-negative scaling factors $y_1,\ldots,y_m$, we can derive the inequality

$$y_1 \cdot (a_{1,1} x_1 + \ldots + a_1 x_n)$$

$$+ \cdots$$

$$+ y_m \cdot (a_{m,1} x_1 + \ldots + a_{m,n} x_n)$$

$$\leq y_1 b_1 + \cdots y_m b_m$$

which is true for every feasible solution $(x_1,\ldots,x_n)$ to the linear program (2). We can rewrite the inequality as

$$(a_{1,1} y_1 + \cdots a_{m,1} y_m) \cdot x_1$$

$$+ \cdots$$

$$+ (a_{1,n} y_1 \cdots a_{m,n} y_m) \cdot x_n$$

$$\leq y_1 b_1 + \cdots y_m b_m$$

So we get that a certain linear function of the x_i is always at most a certain value, for every feasible $(x_1,\ldots,x_n)$. The trick is now to choose the y_i so that the linear function of the x_i for which we get an upper bound is, in turn, an upper bound to the cost function of $(x_1,\ldots,x_n)$. We can achieve this if we choose the y_i such that

Now we see that for every non-negative $(y_1,\ldots,y_m)$ that satisfies, and for every $(x_1,\ldots,x_n)$ that is feasible for,

$$c_1 x_1 + \ldots c_n x_n$$
$$\leq (a_{1,1} y_1 + \cdots a_{m,1} y_m) \cdot x_1$$
$$+ \cdots$$
$$+ (a_{1,n} y_1 \cdots a_{m,n} y_m) \cdot x_n$$
$$\leq y_1 b_1 + \cdots y_m b_m$$

Clearly, we want to find the non-negative values $y_1, \ldots, y_m$ such that the above upper bound is as strong as possible, that is we want to

$$\begin{aligned} &\text{minimize} && b_1 y_1 + \cdots b_m y_m \\ &\text{subject to} && \\ &&& a_{1,1} y_1 + \cdots a_{m,1} y_m \geq c_1 \\ &&& \vdots \\ &&& a_{n,1} y_1 + \ldots + a_{m,n} y_m \geq c_n \\ &&& y_1 \geq 0 \\ &&& \vdots \\ &&& y_m \geq 0 \end{aligned}$$

So we find out that if we want to find the scaling factors that give us the best possible upper bound to the optimum of a linear program in standard maximization form, we end up with a new linear program, in standard minimization form.

Definition 1 If

$$\begin{aligned} &\text{maximize} && c^T x \\ &\text{subject to} && \\ &&& A_x \leq b \\ &&& x \geq 0 \end{aligned}$$

is a linear program in maximization standard form, then its dual is the minimization linear program

$$\begin{array}{ll} \text{minimize} & b^T y \\ \text{subject to} & \\ & A^T y \geq c \\ & y \geq 0 \end{array}$$

So if we have a linear program in maximization linear form, which we are going to call the *primal* linear program, its dual is formed by having one variable for each constraint of the primal (not counting the non-negativity constraints of the primal variables), and having one constraint for each variable of the primal (plus the non-negative constraints of the dual variables); we change maximization to minimization, we switch the roles of the coefficients of the objective function and of the right-hand sides of the inequalities, and we take the transpose of the matrix of coefficients of the left-hand side of the inequalities.

The optimum of the dual is now an upper bound to the optimum of the primal. How do we do the same thing but starting from a minimization linear program? We can rewrite

$$\begin{array}{ll} \text{minimize} & c^T y \\ \text{subject to} & \\ & Ay \geq b \\ & y \geq 0 \end{array}$$

in an equivalent way as

$$\begin{array}{ll} \text{maximize} & -c^T y \\ \text{subject to} & \\ & -Ay \leq -b \\ & y \leq 0 \end{array}$$

If we compute the dual of the above program we get

$$\begin{array}{ll} \text{minimize} & -b^T z \\ \text{subject to} & \\ & -A^T z \geq -c \\ & z \geq 0 \end{array}$$

that is,

maximize $b^T z$

subject to

$$A^T z \leq c$$

$$y \geq 0$$

So we can form the dual of a linear program in minimization normal form in the same way in which we formed the dual in the maximization case:

- Switch the type of optimization,
- Introduce as many dual variables as the number of primal constraints (not counting the non-negativity constraints),
- Define as many dual constraints (not counting the non-negativity constraints) as the number of primal variables.
- Take the transpose of the matrix of coefficients of the left-hand side of the inequality,
- Switch the roles of the vector of coefficients in the objective function and the vector of right-hand sides in the inequalities.

Note that:

***Fact* 2** The dual of the dual of a linear program is the linear program itself.

We have already proved the following:

***Fact* 3** If the primal (in maximization standard form) and the dual (in minimization standard form) are both feasible, then

opt(primal) ≤ opt(dual)

Which we can generalize a little

Theorem 4 (Weak Duality Theorem) If LP_1 is a linear program in maximization standard form, LP_1 is a linear program in minimization standard form, and LP_1 and LP_2 are duals of each other then:

- If LP_1 is unbounded, then LP_2 is infeasible;
- If LP_2 is unbounded, then LP_1 is infeasible;
- If LP_1 and LP_2 are both feasible and bounded, then

  $opt(LP_1) \leq opt(LP_2)$

Proof: We have proved the third statement already. Now observe that the third statement is also saying that if LP_1 and LP_2 are both feasible, then they have to both be bounded, because every feasible solution to LP_2 gives a finite upper bound to the optimum of LP_1 (which then cannot be $+\infty$) and every feasible solution to LP_1 gives a finite lower bound to the optimum of LP_2 (which then cannot be $-\infty$).

What is surprising is that, for bounded and feasible linear programs, there is always a dual solution that certifies the exact value of the optimum.

Theorem 5 (Strong Duality) If either LP_1 or LP_2 is feasible and bounded, then so is the other, and $opt(LP_1) = opt(LP_2)$

To summarize, the following cases can arise:

- If one of LP_1 or LP_2 is feasible and bounded, then so is the other;
- If one of LP_1 or LP_2 is unbounded, then the other is infeasible;
- If one of LP_1 or LP_2 is infeasible, then the other cannot be feasible and bounded, that is, the other is going to be either infeasible or unbounded. Either case can happen.

3.3.2 Konig's Theorem

In the mathematical area of graph theory, Konig's theorem describes an equivalence between the *maximum matching* problem and the *minimum vertex cover* problem in *bipartite graphs*.

Given a bipartite graph G=(V,E)G=(V,E): min |C| = max |M|

We can break the proof into two steps. Firstly, we can prove that |C| ≥≥ |M|, and secondly, we prove that min|C| ≤≤ max|M|, then Konig's theorem gets proved.

It is very easy to prove that |C| ≥≥ |M| for any vertex cover an matching in the same bipartite graph. Because each edge of the matching must be covered by the vertex cover, so at least one vertex of each edge must in the set of vertex cover, thus we proved that |C| ≥≥ |M| at any circumstance.

In this step, we just need to prove that min|C| ≤≤ |M| for some matching M. Given a minimal vertex cover C, if we can extend it into a matching, then we gets proved.

If we convert maximum matching problem and minimum vertex cover problem into Linear Programming problems, then we can prove Konig's theorem elegantly by using duality.

Given a bipartite graph G=(V,E)G=(V,E), the maximum matching can be defined as:

max∑e∈Exemax∑e∈Exe

s.t.

∑e∈δ(v)xe≤1∑e∈δ(v)xe≤1

xe≥0xe≥0

The minimum vertex cover can be defined as:

min∑v∈Gyvmin∑v∈Gyv

s.t.

$\sum(u,v)\in Eyu+yv\geq 1\sum(u,v)\in Eyu+yv\geq 1$

$yv\geq 0yv\geq 0$

We can easily rewrite the above two linear programming problems into primary-dual format. First, we can define a matrix A, where A is a |V| ×× |E| matrix, $Aij=1Aij=1$ if verteix i is an end point of edge j, otherwise, $Aij=0Aij=0$. Then the maximum matching can be defined as:

$\max 1TX\max 1TX$

s.t.

$AX\leq 1AX\leq 1$

$X\geq 0X\geq 0$

And the minimum vertex cover can be defined as:

$\min Y1T\min Y1T$

s.t.

$YTA\geq 1YTA\geq 1$

$Y\geq 0Y\geq 0$

Due to the primary-dual property of linear programming, we can certainly say that Konig's theorem gets proved.

3.3.3 Spanning Trees

A *spanning tree* of a graph is just a subgraph that contains all the vertices and is a tree. A graph may have many spanning trees; for instance the complete graph on four vertices

```
o---o
|\ /|
| X |
|/ \|
o---o
```

has sixteen spanning trees:

```
o---o    o---o    o   o    o---o
|   |    |        |   |    |
|   |    |        |   |    |
|   |    |        |   |    |
o   o    o---o    o---o    o---o
```

```
o---o     o   o     o   o     o   o
 \ /      |\ /       \ /       \ /|
  X       | X         X         X |
 / \      |/ \       / \       / \|
o   o     o   o     o---o     o   o

o   o     o---o     o   o     o---o
|\  |        /      |  /|      \
| \ |       /       | / |       \
|  \|      /        |/  |        \
o   o     o---o     o   o     o---o

o---o     o   o     o   o     o---o
|\        |  /       \  |        /|
| \       | /         \ |       / |
|  \      |/           \|      /  |
o   o     o---o     o---o     o   o
```

➢ Sometimes, you are given graph, but only need one way to connect nodes.

- For example, a network with redundant connections.
- When routing data, you don't care about redundancy: just need a way to send the data to its destination.
- So if we are looking at routing decisions, we basically want to reduce the graph to a tree. The tree tells us exactly how we will send the data.
- If a link fails, we can choose a different tree.

➢ Given a graph $G=(V,E)$, a subgraph of G that is connects all of the vertices and is a tree is called a spanning tree.

- For example, suppose we start with this graph:

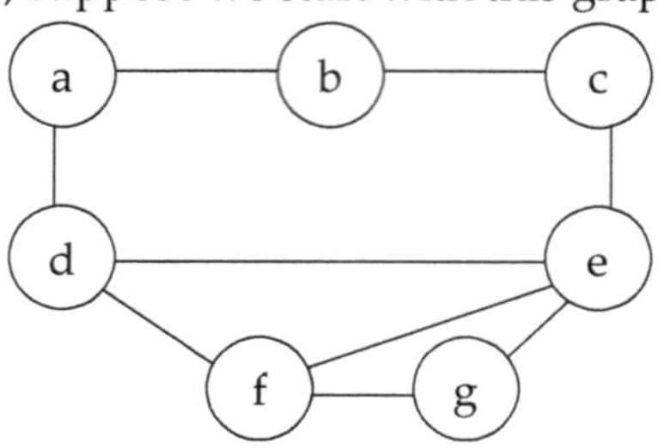

- We can remove edges until we are left with a tree: the result is a spanning tree.

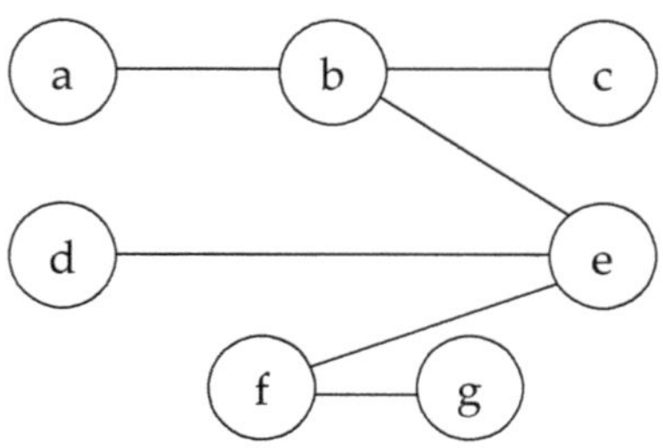

- Clearly, a spanning tree will have $|V|-1$ edges, like any other tree.

➢ Built into Ethernet, there is a Spanning Tree Protocol.

- The job of STP is to take a network with redundant connections and build a spanning tree.
- ... which can be used to send data without worrying about forwarding data around loops and bringing down the network.

➢ **Theorem:** A graph is connected iff it has a spanning tree.

Proof: If a graph is connected, we can identify a cycle and remove an edge from it: it will still be connected. We can continue this until no cycles remain. The result is a spanning tree.

If we have a graph with a spanning tree, then every pair of vertices is connected in the tree. Since the spanning tree is a subgraph of the original graph, the vertices were connected in the original as well.

Minimum Spanning Trees

➢ If we just want a spanning tree, any $n-1$ edges will do.

- If we have edge weights, we can ask for the spanning tree with the lowest total edge weights.
- This is the minimum spanning tree.

➢ For example, if we add some edge weights, what is the minimum spanning tree?

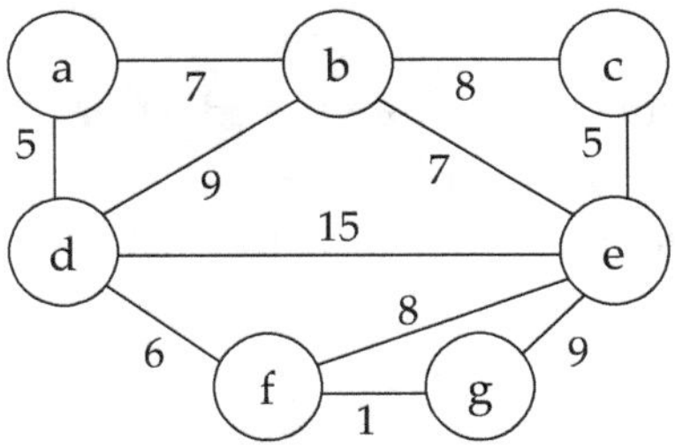

- Is there a general algorithm to find the MST?
- Actually there are two.

➢ Prim's Algorithm works by building a tree from the minimum weight edge, out.

➢ procedure prims_algorithm(V, E)

➢ Et = {minimum weight edge in E}

➢ repeat n-2 times

➢ e = the edge in E that is adjacent to an edge in Et, doesn't form a circuit, and has minimum weight

➢ Et = Et + {e}

return (V, Et)

- That is, we start with the minimum weight edge as our partial-tree.
- Look for an adjacent edge that we can add (no loops) with minimum weight.
- Keep adding edges until we have n−1 of them.

- Kruskal's Algorithm works by building a tree from the minimum weight edge, out.
- procedure kruskals_algorithm(V, E)
- Et = {}
- repeat n-1 times
- e = the edge in E that doesn't form a circuit, and has minimum weight
- Et = Et + {e}

return (V, Et)

- The only difference here is that we don't limit our search to edges near the tree we're building.

- Both of these algorithms find the MST.
- On our example graph with weighted edges…
  - Prim's find the edges of the MST in this order: *fg,df,ad,ab,be,ce*.
  - Kruskal's find the edges in this order: *fg,ad,ce,df,ab,be*.
  - It's the same set of edges (so the same MST), just a difference in how we find them.
- Running time?
  - It depends on how we store the edges and how quickly we can find the minimum-weight ones.
  - But let's assume we're clever about it. Then…
  - Prim's runs in $O(|E|+|V|\log|V|)$.
  - Kruskal's runs in $O(|E|\log|V|)$.
  - So if we have a lot of edges ($|E|=O(|V|2)$), then Prim's will win.
  - If the graph is fairly sparse, Kruskal's will win.
- Both of these are examples of **greedy algorithms**.
  - In a greedy algorithm, you make the choice that seems best right now.
  - *e.g.* the minimum weight edge you can find that doesn't form a circuit.
  - If a greedy algorithm works, it's probably fast, because you can make an easy choice.
  - But it doesn't get the right answer for every problem.
  - *e.g.* travelling salesman: choosing the minimum weight edge will often find a bad overall solution

3.3.4 Greedy Algorithm for Matroids

Greedy algorithms are by far one of the easiest and most well-understood algorithmic techniques. There is a wealth of variations, but at its core the greedy algorithm optimizes something using the natural rule, "pick what looks best" at any step.

Matroids were first introduced by Hassler Whitney, and independently discovered a little later by B.L. van der Waerden. They were both interested in devising a general description of "independence," the properties of which are strikingly similar when specified in linear algebra and graph theory. Since then the study of matroids has blossomed into a large and beautiful theory, one part of which is the characterization of the greedy algorithm: greedy is optimal on a problem if and only if the problem can be represented as a matroid. Mathematicians have also characterized which matroids can be modeled as spanning trees of graphs. As such, matroids have become a standard topic in the theory and practice of algorithms.

Matroid

Let X be a finite set. An *independence system* over X is a family $\mathcal{I}$ of subsets of X with the following two properties.

1. $\mathcal{I}$ is nonempty.
2. If $I \in \mathcal{I}$, then so is every subset of I.

This is too general to characterize greedy algorithms, so we need one more property shared by our examples. There are a few things we do, but here's one nice property that turns out to be enough.

A *matroid* $\mathcal{M} = (X, \mathcal{I})$ is a set X and an independence system $\mathcal{I}$ over X with the following property:

If *A*, *B* are in $\mathcal{I}$ with |A|= |B| + 1, then there is an element $x \in A \setminus B$ such that $B \cup \{a\} \in \mathcal{I}$.

In other words, this property says if I have an independent set that is not *maximally* independent, I can grow the set by adding some suitably-chosen element from a larger independent set. We'll call this the *extension property.* For a warmup exercise, let's prove that the extension property is equivalent to the following (assuming the other properties of a matroid):

For every subset $Y \subset X$, all maximal independent sets *contained in* Y have equal size.

Proof. For one direction, if you have two maximal sets A, B $\subset Y \subset X$ that are not the same size (say *A* is bigger), then you can take any subset of *A* whose size is exactly |B|+1, and use the extension property to make *B* larger, a contradiction. For the other direction, say that I know all maximal independent sets of any $Y \subset X$ have the same size, and you give me $A, B \subset X$. I need to find an $a \in A \setminus B$ that

I can add to B and keep it independent. What I do is take the subset $Y = A \cup B$. Now the sizes of A, B don't change, but B can't be maximal inside Y because it's smaller than A (A might not be maximal either, but it's still independent). And the only way to extend B is by adding something from A, as desired.

So we can use the extension property and the cardinality property interchangeably when talking about matroids. Continuing to connect matroid language to linear algebra and graph theory, the maximal independent sets of a matroid are called *bases,* the size of any basis is the *rank* of the matroid, and the minimal dependent sets are called *circuits*. In fact, you can characterize matroids in terms of the properties of their circuits, which are dual to the properties of bases (and hence all independent sets) in a very concrete sense.

But while you could spend all day characterizing the many kinds of matroids and comatroids out there, we are still faced with the task of seeing how the greedy algorithm performs on a matroid. That is, suppose that your matroid $M = (X, J)$ has a non-negative real number $w(x)$ associated with each $x \in X$. And suppose we had a black-box function to determine if a given set $S \subset X$ is independent. Then the greedy algorithm maintains a set B, and at every step adds a minimum weight element that maintains the independence of B. If we measure the cost of a subset by the sum of the weights of its elements, then the question is whether the greedy algorithm finds a minimum weight basis of the matroid.

Suppose that $M=(X, J)$ is an independence system, and that we have a black-box algorithm to determine whether a given set is independent. Define the greedy algorithm to iteratively adds the cheapest element of X that maintains independence. Then the greedy algorithm produces a maximally independent set S of minimal cost for *every* non-negative cost function on X, if and only if M is a matroid.

It's clear that the algorithm will produce a set that is maximally independent. The only question is whether what it produces has minimum weight among all maximally independent sets. We'll break the theorem into the two directions of the "if and only if":

Part 1: If M is a matroid, then greedy works perfectly no matter the cost function.
Part 2: If greedy works perfectly for every cost function, then M is a matroid.

Call the cost function $w : X \to \mathbb{R}^{\geq 0}$, and suppose that the greedy algorithm picks elements $B = \{x_1, x_2, \ldots, x_r\}$ (in that order). It's easy to see that $w(x_1) \leq w(x_2) \leq \ldots \leq w(x_r)$. Now if you give me *any* list of r independent elements $y_1, y_2, \ldots, y_r \in X$ that has $w(y_1) \leq \ldots \leq w(y_r)$ I claim that $w(x_i) \leq w(y_i)$ for all i. This proves what we want, because if there were a basis of size r with smaller weight, sorting its elements by weight would give a list contradicting this claim.

To prove the claim, suppose to the contrary that it were false, and for some k we have $w(x_i) \leq w(y_i)$. Moreover, pick the smallest k for which this is true. Note $k > 1$, and so we can look at the special sets $S = \{x_1, \ldots, x_{k-1}\}$ and $T = \{y_1, \ldots y_k\}$. Now $|T| = |S| + 1$, so by the matroid property there is some j between 1 and r so that

$S \cup \{y_j\}$ is an independent set (and y_j is not in S). But then $w(y_j) \leq w(y_k) < w(x_k)$, and so the greedy algorithm would have picked y_j before it picks x_k (and the strict inequality means they're different elements). This contradicts how the greedy algorithm runs, and hence proves the claim.

We'll prove this contrapositively as follows. Suppose we have our independence system and it doesn't satisfy the last matroid condition. Then we'll construct a special weight function that causes the greedy algorithm to fail. So let A,B be independent sets with $|A| = |B| + 1$, but for every $a \in A \setminus B$ adding a to B never gives you an independent set.

Now what we'll do is define our weight function so that the greedy algorithm picks the elements we want in the order we want (roughly). In particular, we'll assign all elements of $A \cap B$ a tiny weight we'll call w_1. For elements of $B - A$ we'll use w_2, and for $A - B$ we'll use w_3, with w_2 for everything else. In a more compact notation:

$$w(x) = \begin{cases} w_1 & \text{if } x \in A \cap B \\ w_2 & \text{if } x \in B - A \\ w_3 & \text{if } x \in A - B \\ w_4 & \text{otherwise} \end{cases}$$

We need two things for this weight function to screw up the greedy algorithm. The first is that $w_1 < w_2 < w_3 < w_4$, so that greedy picks the elements in the order we want. Note that this means it'll first pick all of $A \cap B$, and then all of $B - A$, and by assumption it won't be able to pick anything from $A - B$, but since B is assumed to be non-maximal, we have to pick at least one element from $X - (A \cup B)$ and pay w_4 for it.

So the second thing we want is that the cost of doing greedy is worse than picking *any* maximally independent set that contains A (and we know that there has to be some maximal independent set containing A). In other words, if we call m the size of a maximally independent set, we want

$$|A \cap B| w_1 + |B - A| w_2 + (m - |B|) w_4 > |A \cap B| w_1 + |A - B| w_3 + (m + |A|) w_4$$

This can be rearranged (using the fact that $|A| = |B| + 1$) to

$$w_3 > |A - B| w_3 - |B - A| w_2$$

The point here is that the greedy picks too many elements of weight w_4, since if we were to start by taking all of A (instead of all of B), then we could get by with one fewer. That might not be optimal, but it's better than greedy and that's enough for the proof.

So we just need to make w_4 large enough to make this inequality hold, while still maintaining $w_2 < w_3$. There are probably many ways to do this, and here's one. Pick some $0 < \epsilon < 1$, and set

$w_1 = \in/|X|$

$w_2 = \in/|B - A|$

$w_3 = (1 + \in)/|A - B|$

$w_4 = 2$

It's trivial that $w_1 < w_2$ and $w_3 < w_4$. For the rest we need some observations. First, the fact that $|A - B| = |B - A| + 1$ implies that $w_2 < w_3$. Second, both $|A - B|$ and $|B - A|$ are nonempty, since otherwise the second property of independence systems would contradict our assumption that augmenting B with elements of A breaks independence. Using this, we can divide by these quantities to get

$$w_4 = 2 > 1 = \frac{|A-B|(1+\varepsilon)}{|A-B|} - \frac{|B-A|\varepsilon}{|B-A|}$$

As a side note, we proved everything here with respect to *minimizing* the sum of the weights, but one can prove an identical theorem for maximization. The only part that's really different is picking the clever weight function in part 2. In fact, you can convert between the two by defining a new weight function that subtracts the old weights from some fixed number N that is larger than any of the original weights. So these two problems really are the same thing.

So if you can prove your problem is a matroid then you have an awesome algorithm automatically. And if you run the greedy algorithm for fun and it seems like it works all the time, then that may be hinting that your problem is a matroid. This is one of the best situations one could possibly hope for.

But as usual, there are a few caveats to consider. They are both related to efficiency. The first is the black box algorithm for determining if a set is independent. In a problem like minimum spanning tree or finding independent columns of a matrix, there are polynomial time algorithms for determining independence. These two can both be done, for example, with Gaussian elimination. But there's nothing to stop our favorite matroid from requiring an exponential amount of time to check if a set is independent. This makes greedy all but useless, since we need to check for independence many times in every round.

Another, perhaps subtler, issue is that the size of the ground set X might be exponentially larger than the rank of the matroid. In other words, at every step our greedy algorithm needs to find a new element to add to the set it's building up. But there could be such a huge ocean of candidates, all but a few of which break independence. In practice an algorithm might be working with X implicitly, so we could still hope to solve the problem if we had enough knowledge to speed up the search for a new element.

There are still other concerns. For example, a naive approach to implementing greedy takes quadratic time, since you may have to look through every element of X to find the minimum-cost guy to add. What if you just have to have faster runtime than $O(n^2)$?

Greedy Algorithm

Our first example is the problem of scheduling a resource among several competing activities. We shall find that a greedy algorithm provides an elegant and simple method for selecting a maximum-size set of mutually compatible activities.

Suppose we have a set $S = \{1, 2, \ldots, n\}$ of n proposed ***activities*** that wish to use a resource, such as a lecture hall, which can be used by only one activity at a time. Each activity i has a ***start time*** s_i and a ***finish time*** $\hat{a}_i$, where $s_i \leq \hat{a}_i$. If selected, activity i takes place during the half-open time interval $[s_i, \hat{a}_i)$. Activities i and j are ***compatible*** if the intervals $[s_i, \hat{a}_i)$ and $[s_j, \hat{a}_j)$ do not overlap (*i.e.*, i and j are compatible if $s_i \geq \hat{a}_j$ or $s_j \geq \hat{a}_i$). The ***activity-selection problem*** is to select a maximum-size set of mutually compatible activities.

A greedy algorithm for the activity-selection problem is given in the following pseudocode. We assume that the input activities are in order by increasing finishing time:

$$\hat{a}_1 \leq \hat{a}_2 \leq \ldots \leq \hat{a}_n .$$

If not, we can sort them into this order in time $O(n \text{ 1g } n)$, breaking ties arbitrarily. The pseudocode assumes that inputs s and â are represented as arrays.

GREEDY-ACTIVITY-SELECTOR(s, f)

1. $n \leftarrow length[s]$
2. $A \leftarrow \{1\}$
3. $j \leftarrow 1$
4. **for** $i \leftarrow 2$ **to** n
5. **do if** $s_i \geq \hat{a}_j$
6. **then** $A \leftarrow A \cup \{i\}$
7. $j \leftarrow i$
8. **return** A

The operation of the algorithm is shown in Figure 3.1. The set A collects the selected activities. The variable j specifies the most recent addition to A. Since the activities are considered in order of nondecreasing finishing time, f_j is always the maximum finishing time of any activity in A. That is,

$$\hat{a}_j = \max\{f_k : K \in A\}.$$

Lines 2-3 select activity 1, initialize A to contain just this activity, and initialize j to this activity. Lines 4-7 consider each activity i in turn and add i to A if it is compatible with all previously selected activities. To see if activity i is compatible with every activity currently in A, it suffices by equation (17.2) to check (line 5) that its start time s_i is not earlier than the finish time f_j of the activity most recently added to A. If activity i is compatible, then lines 6-7 add it to A and update j. The GREEDY-ACTIVITY-SELECTOR procedure is quite efficient. It can schedule a set S of n activities in $\Theta(n)$ time, assuming that the activities were already sorted initially by their finish times.

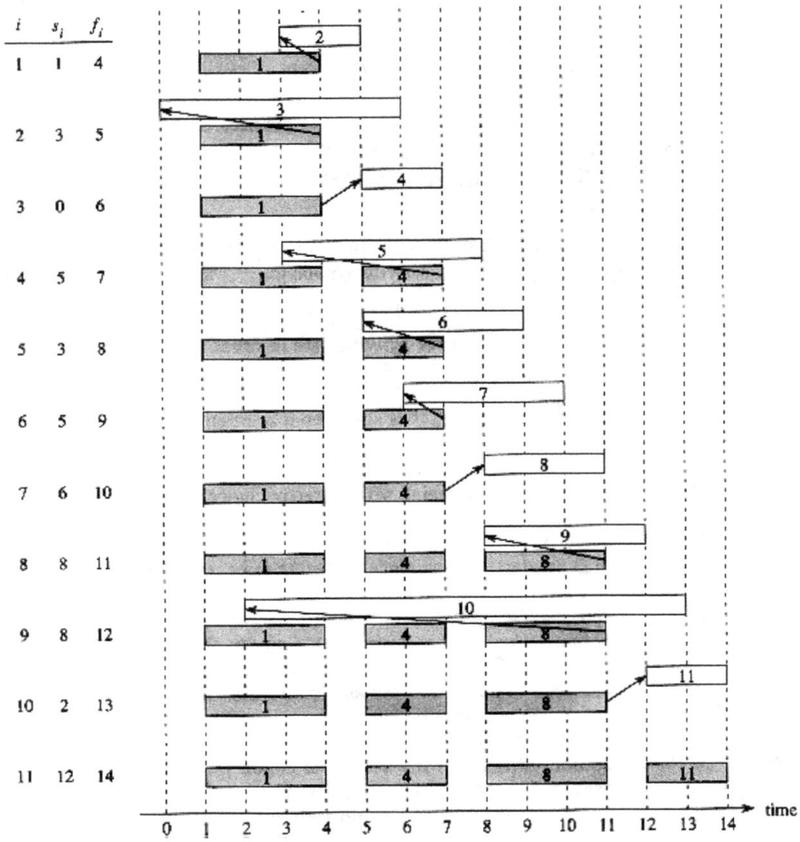

Figure 3.1: Greedy-Activity-Selector.

The activity picked next by GREEDY-ACTIVITY-SELECTOR is always the one with the earliest finish time that can be legally seheduled. The activity picked is thus a "greedy" choice in the sense that, intuitively, it leaves as much opportunity as possible for the remaining activities to be scheduled. That is, the greedy choice is the one that maximizes the amount of unscheduled time remaining.

Proving the Greedy Algorithm Correct

Greedy algorithms do not always produce optimal solutions. However, GREEDY-ACTIVITY-SELECTOR always finds an optimal solution to an instance of the activity-selection problem.

Algorithm GREEDY-ACTIVITY-SELECTOR produces solutions of maximum size for the activity-selection problem.

Let $S = \{1, 2, \ldots, n\}$ be the set of activities to schedule. Since we are assuming that the activities are in order by finish time, activity 1 has the earliest finish time.

Suppose that $A \subseteq S$ is an optimal solution to the given instance of the activity-selection problem, and let us order the activities in A by finish time. Suppose further that the first activity in A is activity k. If $k = 1$, then schedule A begins with a greedy choice. If $k \neq 1$, we want to show that there is another optimal solution B to S that begins with the greedy choice, activity 1. Let $B = A - \{k\} \cup \{1\}$. Because $f_i \leq f_k$, the activities in B are disjoint, and since B has the same number of activities as A, it is also optimal. Thus, B is an optimal solution for S that contains the greedy choice of activity 1. Therefore, we have shown that there always exists an optimal schedule that begins with a greedy choice.

Moreover, once the greedy choice of activity 1 is made, the problem reduces to finding an optimal solution for the activity-selection problem over those activities in S that are compatible with activity 1. That is, if A is an optimal solution to the original problem S, then $A' = A - \{1\}$ is an optimal solution to the activity-selection problem $S' = \{i \in S: S_i \geq f_1\}$. Why? If we could find a solution B' *to* S′ with more activities than A', *adding activity 1 to* B′ would yield a solution B to S with more activities than A, thereby contradicting the optimality of A. Therefore, after each greedy choice is made, we are left with an optimization problem of the same form as the original problem. By induction on the number of choices made, making the greedy choice at every step produces an optimal solution.

Greedy-Choice Property

The first key ingredient is the ***greedy-choice property***: a globally optimal solution can be arrived at by making a locally optimal (greedy) choice. Here is where greedy algorithms differ from dynamic programming. In dynamic programming, we make a choice at each step, but the choice may depend on the solutions to subproblems. In a greedy algorithm, we make whatever choice seems best at the moment and then solve the subproblems arising after the choice is made. The choice made by a greedy algorithm may depend on choices so far, but it cannot depend on any future choices or on the solutions to subproblems. Thus, unlike dynamic programming, which solves the subproblems bottom up, a greedy strategy usually progresses in a top-down fashion, making one greedy choice after another, iteratively reducing each given problem instance to a smaller one.

Of course, we must prove that a greedy choice at each step yields a globally optimal solution, and this is where cleverness may be required. Typically, as in the case of Theorem, the proof examines a globally optimal solution. It then shows that the solution can be modified so that a greedy choice is made as the first step, and that this choice reduces the problem to a similar but smaller problem. Then, induction is applied to show that a greedy choice can be used at every step. Showing that a greedy choice results in a similar but smaller problem reduces the proof of correctness to demonstrating that an optimal solution must exhibit optimal substructure.

Optimal Substructure

A problem exhibits ***optimal substructure*** if an optimal solution to the problem contains within it optimal solutions to subproblems. This property is a key ingredient of assessing the applicability of dynamic programming as well as greedy algorithms. As an example of optimal substructure, recall that the proof of Theorem demonstrated that if an optimal solution A to the activity selection problem begins with activity 1, then the set of activities $A' = A - \{1\}$ is an optimal solution to the activity-selection problem $S' = \{i \in S : s_i \geq \hat{a}_1\}$.

Greedy versus Dynamic Programming

Because the optimal-substructure property is exploited by both greedy and dynamic-programming strategies, one might be tempted to generate a dynamic-programming solution to a problem when a greedy solution suffices, or one might mistakenly think that a greedy solution works when in fact a dynamic-programming solution is required. To illustrate the subtleties between the two techniques, let us investigate two variants of a classical optimization problem.

The ***0-1 knapsack problem*** is posed as follows. A thief robbing a store finds n items; the ith item is worth v_i dollars and weighs w_i pounds, where v_i and w_i are integers. He wants to take as valuable a load as possible, but he can carry at most W pounds in his knapsack for some integer W.

In the ***fractional knapsack problem,*** the setup is the same, but the thief can take fractions of items, rather than having to make a binary (0-1) choice for each item. You can think of an item in the 0-1 knapsack problem as being like a gold ingot, while an item in the fractional knapsack problem is more like gold dust.

Both knapsack problems exhibit the optimal-substructure property. For the 0-1 problem, consider the most valuable load that weighs at most W pounds. If we remove item j from this load, the remaining load must be the most valuable load weighing at most $W - w_j$ that the thief can take from the $n - 1$ original items excluding j. For the comparable fractional problem, consider that if we remove a weight w of one item j from the optimal load, the remaining load must be the most valuable load weighing at most $W - w$ that the thief can take from the $n - 1$ original items plus $w_j - w$ pounds of item j.

Although the problems are similar, the fractional knapsack problem is solvable by a greedy strategy, whereas the 0-1 problem is not. To solve the fractional problem, we first compute the value per pound v_i/w_i for each item. Obeying a greedy strategy, the thief begins by taking as much as possible of the item with the greatest value per pound. If the supply of that item is exhausted and he can still carry more, he takes as much as possible of the item with the next greatest value per pound, and so forth until he can't carry any more. Thus, by sorting the items by value per pound, the greedy algorithm runs in $O(n\lg n)$ time.

To see that this greedy strategy does not work for the 0-1 knapsack problem, consider the problem instance illustrated in Figure 3.2. There are 3 items, and the knapsack can hold 50 pounds. Item 1 weighs 10 pounds and is worth 60 dollars. Item 2 weighs 20 pounds and is worth 100 dollars. Item 3 weighs 30 pounds and is worth 120 dollars. Thus, the value per pound of item 1 is 6 dollars per pound, which is greater than the value per pound of either item 2 (5 dollars per pound) or item 3 (4 dollars per pound). The greedy strategy, therefore, would take item 1 first. As can be seen from the case analysis in Figure 3.2, however, the optimal solution takes items 2 and 3, leaving 1 behind. The two possible solutions that involve item 1 are both suboptimal.

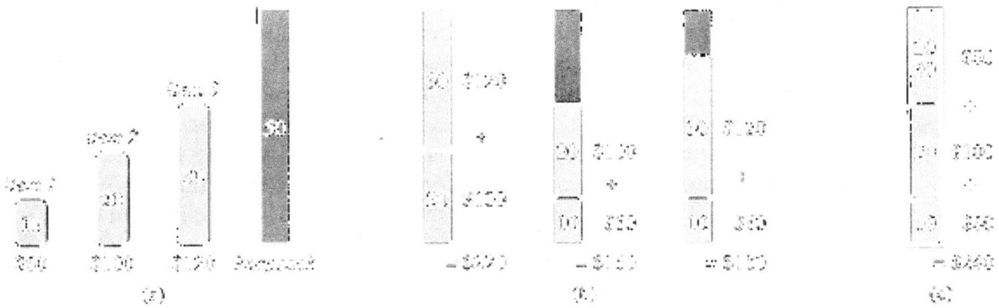

Figure 3.2: Greedy Strategy.

3.3.5 Greedy Algorithms on a Weighted Matroid

Many problems for which a greedy approach provides optimal solutions can be formulated in terms of finding a maximum-weight independent subset in a weighted matroid. That is, we are given a weighted matroid $M = (S, T)$, and we wish to find an independent set $A \in I$ such that $w(A)$ is maximized. We call such a subset that is independent and has maximum possible weight an ***optimal*** subset of the matroid. Because the weight $w(x)$ of any element $x \in S$ is positive, an optimal subset is always a maximal independent subset – it always helps to make A as large as possible.

For example, in the ***minimum-spanning-tree problem***, we are given a connected undirected graph $G = (V, E)$ and a length function w such that $w(e)$ is the (positive) length of edge e. (We use the term "length" here to refer to the original edge weights for the graph, reserving the term "weight" to refer to the weights in the associated matroid.) We are asked to find a subset of the edges that connects all of the vertices together and has minimum total length. To view this as a problem of finding an optimal subset of a matroid, consider the weighted matroid M_G with weight function w', *where* $w'(e) = w_0 - w(e)$ and w_0 is larger than the maximum length of any edge. In this weighted matroid, all weights are positive and an optimal subset is a spanning tree of minimum total length in the original graph. More specifically, each maximal independent subset A corresponds to a spanning tree, and since

$$w'(A) = (|V| - 1)w_0 - w(A)$$

for any maximal independent subset A, the independent subset that maximizes $w'(A)$ must minimize $w(A)$. Thus, any algorithm that can find an optimal subset A in an arbitrary matroid can solve the minimum-spanning-tree problem.

The algorithm takes as input a weighted matroid $M = (S, \mathcal{I})$ with an an associated positive weight function w, and it returns an optimal subset A. In our pseudocode, we denote the components of M by $S[M]$ and $\mathcal{I}[M]$ and the weight function by w. The algorithm is greedy because it considers each element $x \in S$ in turn in order of nonincreasing weight and immediately adds it to the set A being accumulated if $A \cup \{x\}$ is independent.

Greedy (*M*, *w*)

1. A $\leftarrow \emptyset$
2. sort $S[M]$ into nonincreasing order by weight w
3. for each $x \in S[\mathrm{x}]$, taken in nonincreasing order by weight $w(x)$
4. do if A $\cup \{x\} \in I\,[M]$
5. then $A \leftarrow A \cup \{x\}$
6. return A

The elements of S are considered in turn, in order of nonincreasing weight. If the element x being considered can be added to A while maintaining A's independence, it is. Otherwise, x is discarded. Since the empty set is independent by the definition of a matroid, and since x is only added to A if $A \cup \{x\}$ is independent, the subset A is always independent, by induction. Therefore, GREEDY always returns an independent subset A. We shall see in a moment that A is a subset of maximum possible weight, so that A is an optimal subset.

The running time of GREEDY is easy to analyze. Let n denote $|S|$. The sorting phase of GREEDY takes time $O(n \lg n)$. Line 4 is executed exactly n times, once for each element of S. Each execution of line 4 requires a check on whether or not the set $A \cup \{x\}$ is independent. If each such check takes time $O(f(n))$, the entire algorithm runs in time $O(n \lg n + nf(n))$.

We now prove that GREEDY returns an optima1 subset.

Lemma

Suppose that $M = (S,I)$ is a weighted matroid with weight function w and that S is sorted into nonincreasing order by weight. Let x be the first element of S such that $\{x\}$ is independent, if any such x exists. If x exists, then there exists an optimal subset A of S that contains x.

If no such x exists, then the only independent subset is the empty set and we're done. Otherwise, let B be any nonempty optimal subset. Assume that $x \notin B$; otherwise, we let $A = B$ and we're done.

No element of B has weight greater than $w(x)$. To see this, observe that $y \in B$ implies that $\{y\}$ is independent, since $B \in I$ *and* I and I is hereditary. Our choice of x therefore, ensures that $w(x) \geq w(y)$ for any $y \in B$.

Construct the set A as follows. Begin with $A = \{x\}$. By the choice of x, A is independent. Using the exchange property, repeatedly find a new element of B that can be added to A until $|A| = |B|$ while preserving the independence of A. Then, $A = B - \{y\} \cup \{x\}$ for some $y \in B$, and so

$$w(A) = w(B) - w(y) + w(x)$$
$$\geq w(B)\ .$$

Because B is optimal, A must also be optimal, and because $x \in A$, the lemma is proven. We next show that if an element is not an option initially, then it cannot be an option later.

Lemma

Let $M = (S,I)$ be any matroid. If x is an element of S such that x is not an extension of $\emptyset$, then x is not an extension of any independent subset A of S.

The proof is by contradiction. Assume that x is an extension of A but not of $\emptyset$. Since x is an extension of A, we have that $A \cup \{x\}$ is independent. Since I is hereditary, $\{x\}$ must be independent, which contradicts the assumption that x is not an extension of $\emptyset$.

Lemma above says that any element that cannot be used immediately can never be used. Therefore, GREEDY cannot make an error by passing over any initial elements in S that are not an extension of $\emptyset$, since they can never be used.

Lemma

Let x be the first element of S chosen by GREEDY for the weighted matroid $M = (S,I)$. The remaining problem of finding a maximum-weight independent subset containing x reduces to finding a maximum-weight independent subset of the weighted matroid $M = (S,I)$, where

$$S' = \{y \in S : \{x,y\} \in I\},$$
$$I' = \{B \subseteq S - \{x\} : B \cup \{x\} \in I\},\ \text{and}$$

the weight function for M' *is the weight function for* M, *restricted to* S'. (We call M' *the* **contraction** *of* M *by the element* x.)

If A is any maximum-weight independent subset of M containing x, then $A' =$ A - *{x} is an independent subset of* M'. Conversely, any independent subset A' *of* M' yields an independent subset $A = A' \cup$ *{x} of* M. *Since we have in both cases that* w(A) = w(A') + $w(x)$, a maximum-weight solution in M containing x yields a maximum-weight solution in M', and vice versa.

Theorem

If $M = (S,I)$ is a weighted matroid with weight function w, then the call GREEDY(M, w) returns an optimal subset.

By above Lemma, any elements that are passed over initially because they are not extensions of $\emptyset$ can be forgotten about, since they can never be useful. Once the first element x is selected, Lemma implies that GREEDY does not err by adding x to A, since there exists an optimal subset containing x. Finally, Lemma 17.9 implies that the remaining problem is one of finding an optimal subset in the matroid M' that is the contraction of M by x. After the procedure GREEDY sets A to$\{x\}$, all of its remaining steps can be interpreted as acting in the matroid '$M' = (S',I')$, because B is independent in M' if and only if $B \cup \{x\}$ is independent in M, for all sets $B \in I'$. Thus, the subsequent operation of GREEDY will find a maximum-weight independent subset for M', and the overall operation of GREEDY will find a maximum-weight independent subset for M.

UNIT 4

Approximation Algorithms

4.1 Introduction

An approximate algorithm is a way of dealing with NP-completeness for optimization problem. This technique does not guarantee the best solution. The goal of an approximation algorithm is to come as close as possible to the optimum value in a reasonable amount of time which is at most polynomial time.

Suppose we have some optimization problem instance i, which has a large number of feasible solutions. Also let $c(i)$ be the cost of solution produced by approximate algorithm and $c^*(i)$ be the cost of optimal solution. For minimization problem, we are interested in finding a solution of a given instance i in the set of feasible solutions, such that $c(i)/c^*(i)$ be as small as possible. On the other hand, for maximization problem, we are interested in finding a solution in the feasible solution set such that $c^*(i)/c(i)$ be as small as possible.

We say that an approximation algorithm for the given problem instance i, has a ratio bound of $p(n)$ if for any input of sign n, the cost c of the solution produced by the approximation algorithm is within a factor of $p(n)$ of the cost c^* of an optimal solution. That is $\max(c(i)/c^*(i), c^*(i)/c(i)) \leq p(n)$

This definition applies for both minimization and maximization problems. Note that $p(n)$ is always greater than or equal to 1. If solution produced by approximation algorithm is true optimal solution then clearly we have $p(n) = 1$.

For a minimization problem, $0 < c^*(i) \leq c(i)$, and the ratio $c(i)/c^*(i)$ gives the factor by which the cost of the approximate solution is larger than the cost of an optimal solution. Similarly, for a maximization problem, $0 < c(i) \leq c^*(i)$, and the ratio $c^*(i)/c(i)$ gives the factor by which the cost of an optimal solution is larger than the cost of the approximate solution.

4.2 Relative Error

We define the relative error of the approximate algorithm for any input size as $\mathrm{mod}[c(i) - c^*(i)/ c^*(i)]$

We say that an approximate algorithm has a relative bound of $\varepsilon(n)$ if $\mathrm{mod}[c(i) - c^*(i)/ c^*(i)] \leq \varepsilon(n)$

4.3 Accuracy Ratio

Let $f(s_a)$ be the value of the objective function, f, of the approximation algorithm and $f(s^*)$ is the value of the exact solution then the **accuracy ratio** is

$r(s_a) = f(s_a)/f(s^*)$ for minimizing the objective function

or $= f(s^*)/f(s_a)$ maximizing the objective function

The approximation called is called c approximate if

$r(s_a) \leq c$

The smallest c that holds for all instances of the problem is called the **performance ratio**.

4.3.1 Approximation Algorithms for Traveling Salesman

Before we present some polynomial approximate algorithms, consider weather or not it is possible to construct a finite c-approximate algorithm for the traveling salesman problem. The answer is no, unless $P = NP$.

Theorem

If exist a polynomial time c-approximation algorithm for the traveling salesman problem then P = NP

We prove by construction, specifically we use the polynomial time c-approximation algorithm for the traveling salesman problem to construct a polynomial time algorithm for Hamilton circuit problem.

Suppose exists such an algorithm A with constant c then an approximate solution, s_a, will have $f(s_a) \leq cf(s^*)$, where s^* is the optimal solution.

We use algorithm to construct the algorithm to solve the Hamiltonian circuit in polynomial time for any graph.

Let G be an arbitrary graph with n vertices. We map G to a complete weighted graph G' by assigning weight 1 to each edge in G and weight $cn+1$ to edges not in G.

1. If G has a Hamiltonian circuit, its length in G' is n.

   It is an optimal solution to the traveling salesman problem for G'.

   If s_a is an approximate solution obtained for G' using algorithm A then $f(s_a) \leq cn$ by assumption (using $f(s_a) \leq cf(s^*)$).

   Meaning, if G has a Hamiltonian circuit then $f(s_a) \leq cn$. (*)

2. If G does not have a Hamiltonian circuit then the shortest tour in G' must contain at least one edge $cn+1$ and $f(s_a) \geq f(s^*) > cn$.

   Meaning, if G does not have a Hamiltonian circuit then $f(s_a) > cn$. (**)

So if the approximation algorithm for the traveling salesman problem runs in polynomial time we can solve the Hamiltonian circuit decision problem in polynomial time by using inequalities (*) and (**).

We suspect that $P \neq NP$, so we suspect there does not exist a polynomial time approximate algorithm with finite c for the traveling salesman problem.

This does not mean that graphs with restrictions cannot have c-approximations.

4.3.2 Nearest-neighbour Approximation for the Traveling Salesman in a Complete Graph

Step 1: Choose an arbitrary city as the start

Step 2: Repeat until all the cities have been visited, go to the nearest unvisited city

Step 3: Return to the starting city

Note that the graph must be complete to insure that the start city is adjacent to the last city. Illustrate the algorithm on a graph of a box with an X inside. Make the return edge large.

calculate $r(s_a) = f(s^*)/f(s_a)$

Show that by changing the weighting of the return edge, to w get $r(s_a) \sim w$

So, $R_A = \infty$ for the nearest neighbor approximation.

Problem:

Given a complete undirected graph $G=(V, E)$ that has non-negative integer cost $c(u, v)$ associated with each edge (u, v) in E, the problem is to find a Hamiltonian cycle (tour) of G with minimum cost.

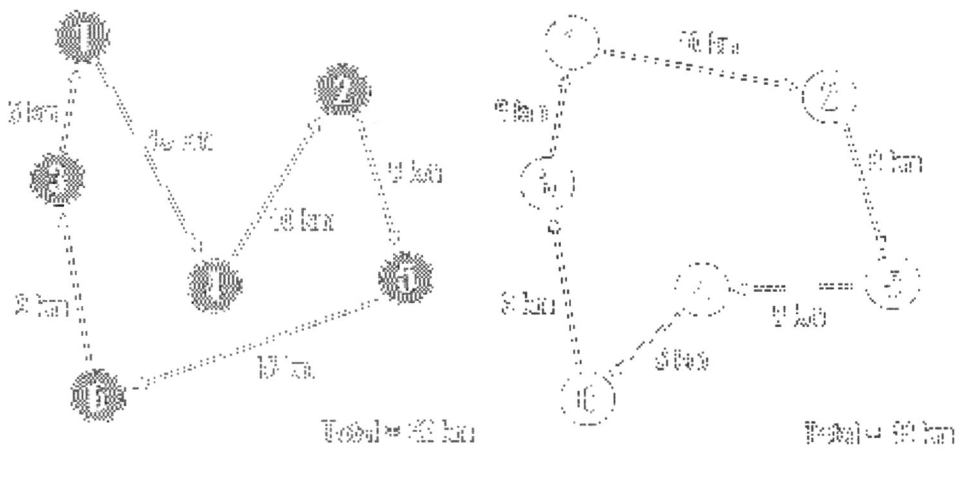

A salespersons starts from the city 1 and has to visit six cities (1 through 6) and must come back to the starting city, *i.e.*, 1. The first route (left side) 1→ 4 → 2 → 5 → 6 → 3 → 1 with the total length of 62 km, is a relevant selection but is not the best solution. The second route (right side) 1 → 2 → 5 → 4 → 6 → 3 → 1 represents the must better solution as the total distance, 48 km, is less than for the first route.

Suppose $c(A)$ denoted the total cost of the edges in the subset A subset of E, *i.e.*,

$$c(A) = \sum_{u,v \text{ in } A} c(u, v)$$

Moreover, the cost function satisfies the triangle inequality. That is, for all vertices u, v, w in V, w have $c(u, w) \leq c(u, v) + c(v, w)$.

Note that the TSP problem is NP-complete even if we require that the cost function satisfies the triangle inequality. This means that it is unlikely that we can find a polynomial-time algorithm for TSP.

TSP with the Triangle-Inequality

When the cost function satisfies the triangle inequality, we can design an approximate algorithm for TSP that returns a tour whose cost is not more than twice the cost of an optimal tour.

Outline of an APPROX-TSP-TOUR

First, compute a MST (minimum spanning tree) whose weight is a lower bound on the length of an optimal TSP tour. Then, use MST to build a tour whose cost is no more than twice that of MST's weight as long as the cost function satisfies triangle inequality.

Operation of APPROX-TSP-TOUR

Let root r be a in following given set of points (graph).

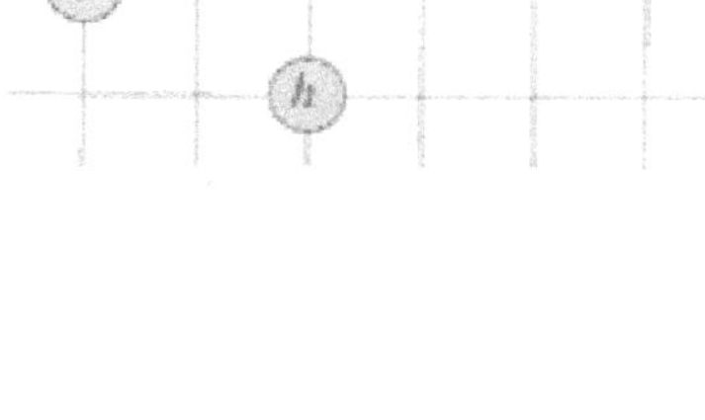

Construct MST from root a using MST-PRIM (G, c, r).

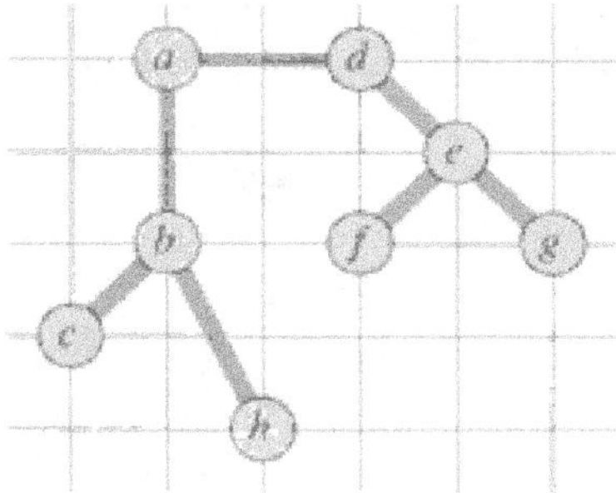

List vertices visited in preorder walk. $L = \{a, b, c, h, d, e, f, g\}$

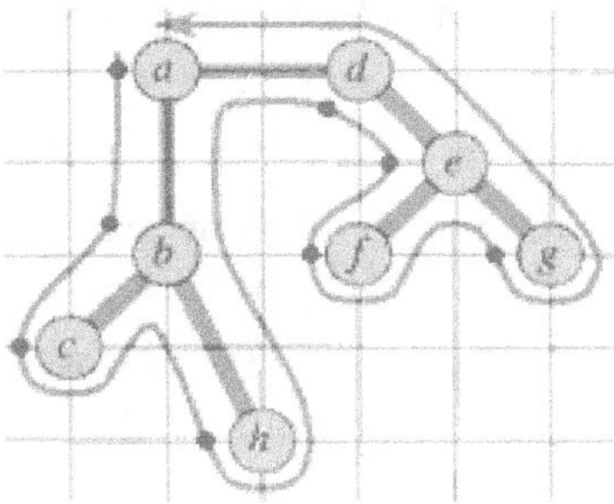

Return Hamiltonian cycle.

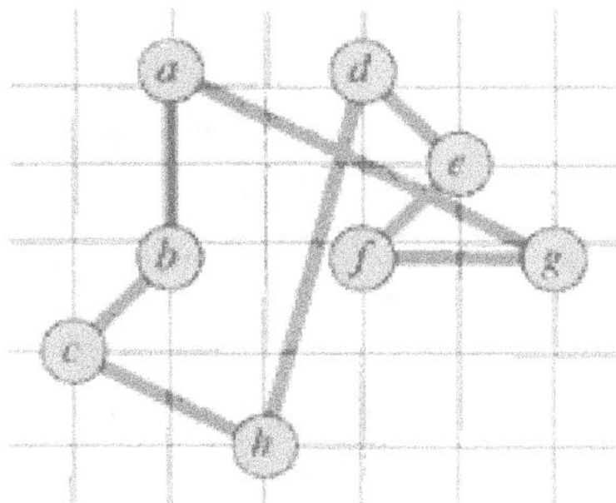

Optimal TSP tour for a given problem (graph) would be

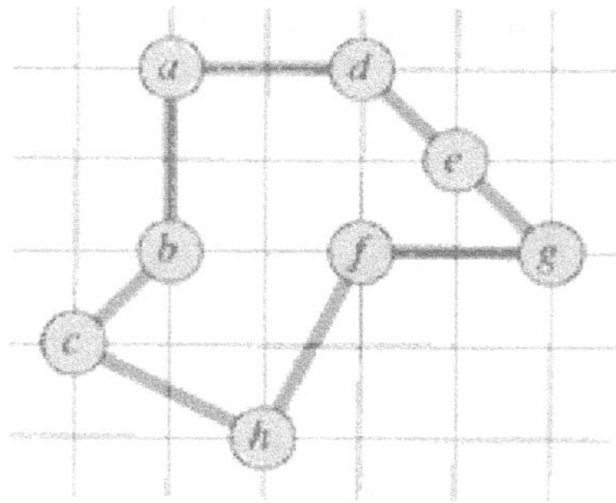

which is about 23% shorter.

Theorem:

APPROX-TSP-TOUR is a polynomial-time 2-approximation algorithm for TSP with triangle inequality.

1. We have already shown that APPROX-TSP-TOUR-time.
2. Let H^* denote the optimal tour. Observe that a TSP with one edge removed is a spanning tree (not necessarily MST).

It implies that the weight of the MST T is in lower bound on the cost of an optimal tour.

$c(T) = c(H^*)$

A "Full" walk, W, traverse every edge of MST, T, exactly twice. That is,

$c(w) = 2c(T)$

which means

$c(w) \le 2c(H^*)$

and we have

$c(w)/c(H^*) \le p(n) = 2$

That is, the cost of walk, $c(w)$, of the solution produced by the algorithm is within a factor of $p(n)=2$ of the cost $c(H^*)$ of an optimal solution.

General TSP

Without the triangle inequality, a polynomial time approximate algorithm with constant approximation ratio not exist unless $P=NP$.

If $P \neq NP$, there is no polynomial-time algorithm with approximation ratio $p\geq1$ for general TSP.

Suppose to the contrary that there exists a polynomial-time algorithm A for $p\geq1$.

Now, use algorithm A to solve instance of HAM-CYCLE problem in polynomial-time. Since HAM-CYCLE problem is NP-complete then by theorem

If any NP-complete problem is polynomial-time solvable, then P=NP. Equivalently, if any problem in NP is not polynomial-time solvable than no NP-complete problem is polynomial solvable.

Solving HAM-CYCLE in polynomial-time implies that *P=NP*.

Since the HAM-CYCLE is NP-complete and we assumed and we assumed that *P≠NP*, a contradiction is arise.

Let $G = (V, E)$ be an instance of Hamiltonian cycle problem. We want to determine efficiently whether G contains a Hamiltonian cycle by making use of an algorithm *A*. We transform G into an instance of the TSP problem as follows:

Consider a complete graph $G` = (V, E`)$ where $E` = \{(u, v): u, v \text{ in } V \text{ and } u \neq v\}$. Assign an integer cost to each edge in *E`* as follows:

$$c(u,v) = \frac{1 \text{ if } (u, v) \text{ in } E}{P|V| + 1 \text{ otherwise}}$$

Because algorithm *A* is guaranteed to return a tour of cost no more than *P* times the cost of an optimal tour, if graph G contains a Hamiltonian cycle, then *A* must return it. If G has no Hamiltonian cycle, then *A* returns a tour of cost more than $P|V|$. Therefore, we can use algorithm *A* to solve Hamiltonian cycle problem in polynomial time and this is impossible unless *P=NP*.

4.3.3 Euclidean Weighted Graph

A Euclidean weighted graph wave weights, $d[i, j]$ with the properties

Triangle inequality: $d[i, j] \leq d[i, k] + d[k, j]$

Symmetry: $d[i, j] = d[j, i]$

The nearest neighbor approximation for the traveling salesman problem in a complete Euclidean graph has accuracy ratio

$f(s_a)/f(s^*) \leq (1/2)(\lg n + 1)$

(given without proof).

Still $R_A = \infty$.

4.3.4 Twice-around the Minimum Spanning Tree

Step 1: Construct a minimum spanning tree of the graph

Step 2: Starting at an arbitrary vertex, walk around the minimum spanning tree recording all the vertices. (This can be done by a tree traversal and recording using preorder and postorder visits.)

Step 3: Scan the vertex list obtained in step 2 and eliminate from it all repeated occurrences of the same vertex except the starting and ending vertex. (This step is equivalent to making short cuts.) The remaining vertices on the list represent the approximate tour.

Again the tree must be complete, for there to exist short cuts.

Theorem 1

The twice around the minimum spanning tree is a 2-approximation algorithm for the traveling salesman with Euclidean distances.

We need to show that $f(s_a) \leq 2f(s^*)$

Removing any edge from s^* yields a spanning tree T with weight $w(T)$.

T is not necessarily a minimum spanning tree (MST) so its weight must be greater than or equal to the weight of a minimum spanning tree, w(MST). Also because removing an edge from s^* reduces $f(s^*)$:

$f(s^*) > w(T) \geq w(MST)$

or

$2f(s^*) > 2w(MST)$ (= length of the walk obtained in step 2) (*)

(above always true)

The possible shortcuts outlined in step 3 cannot increase the length of the tour in a graph with Euclidian distances, so using the definition of the objective function

the length of the walk obtained in step 2 (= $2w(MST)$) ≥ the length of the tour (= $f(s_a)$)

$2w(MST) \geq f(s_a)$ (**)

(above true only for graphs with Euclidian distances)

Combining (*) and (**)

$2f(s^*) > f(s_a)$

4.4 Vertex Cover

A vertex-cover of an undirected graph G=(V, E) is a subset of V`subset of $V$ such that if edge ($u$, $v$) is an edge of $G$ then either $u$ in $V$ or $v$ in $V$` (or both).

Example

Find a vertex-cover of maximum size in a given undirected graph.

This optimal vertex-cover is the optimization version of an NP-complete problem but it is not too hard to find a vertex-cover that is near optimal.

APPROX-VERTEX_COVER (G: Graph)

1. $c \leftarrow \{\}$
2. $E` \leftarrow E[G]$

3. While $E`$ is not empty do
4. Let (u, v) be an arbitrary edge of $E`$
5. $c \leftarrow c \cup \{u, v\}$
6. Remove from $E`$ every edge incident on either u or v
7. Return c

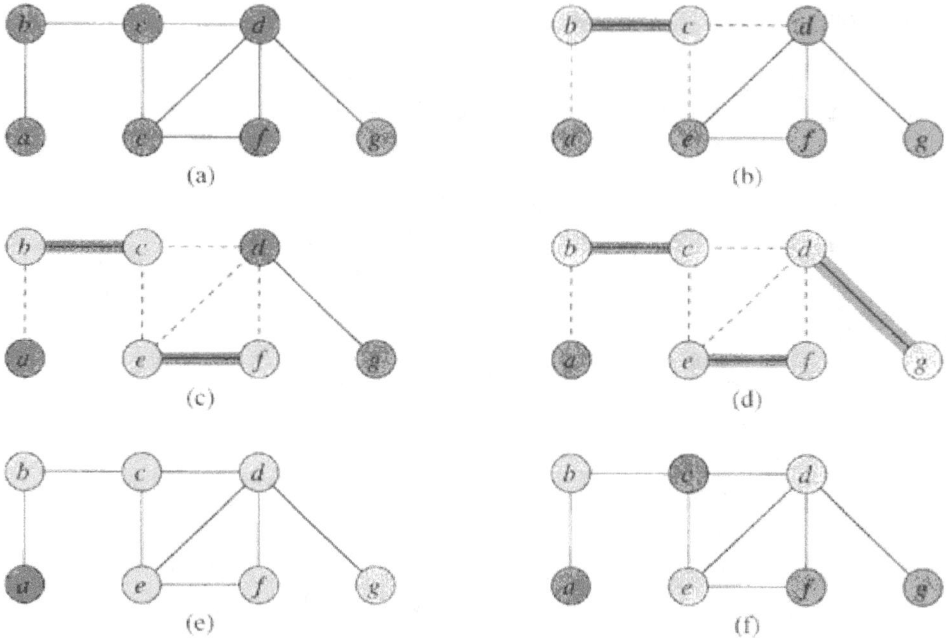

It is easy to see that the running time of this algorithm is $O(V+E)$, using adjacency list to represent $E`$.

APPROX-VERTEX-COVER is a polynomial-time 2-approximate algorithm, i.e., the algorithm has a ration bound of 2.

Since this a minimization problem, we are interested in smallest possible c/c^*. Specifically we want to show $c/c^* \leq 2 = p(n)$. In other words, we want to show that APPROX-VERTEX-COVER algorithm returns a vertex-cover that is atmost twice the size of an optimal cover.

Let the set c and c^* be the sets output by APPROX-VERTEX-COVER and OPTIMAL-VERTEX-COVER respectively. Also, let A be the set of edges selected by line 4.

Because, we have added both vertices, we get $c = 2|A|$ but OPTIMAL-VERTEX-COVER would have add one of two.

=> $c/c^* \leq p(n) = 2$.

Formally, since no two edge in A are covered by the same vertex from c^* (since, once an edge is picked in line 4, all other edges that are incident on its endpoints are deleted from $E`$ in line 6) and we the lower bound:

$$|c^*| \geq A \tag{1}$$

On the size of an OPTIMAL-VERTEX-COVER.

In line 4, we picked both en points yielding an upper bound on the size of Vertex-Cover.

$$|c| \leq 2|A|$$

Since, upper bound is an exact in this case, we have

$$|c| = 2|A| \tag{2}$$

Take $|c|/2 = |A|$ and put it in equation 1

$$|c^*| \geq |c|/2$$

$$|c^*|/|c| \geq 1/2$$

$$|c^*|/|c| \leq 2 = p(n)$$

proving the theorem.

UNIT 5

Integer Programming

5.1 Linear Program

Linear programs are mathematical models and there are algorithms to solve them. Let's start with a linear expression. This is called the *objective function*:

$$obj(x_1, x_2, \ldots, x_n) = a_1 x_1 + a_2 x_2 + \cdots + a_n x_n \text{ each } x_i \geq 0$$

The goal of linear programming is to minimize some objective function. By "some" objective function, it means that n and a's are all known for a particular problem. *e.g.*, 3 * x1 + 14 * x2

Now, of course, if that was all there was to it, we could just set all the x to zero. However, linear programs are also allowed to have one or more *constraints*. Constraints are each simply another linear expression, set less than or equal to zero. Note that the coefficients of the x's can be different than the objective, and there can be a constant term as well. *e.g.*,

$$b_0 + b_1 x_1 + b_2 x_2 + \cdots + b_n x_n \leq 0$$

$$c_0 + c_1 x_1 + c_2 x_2 + \cdots + c_n x_n \leq 0$$

and so on. There can be as many constraints as you like. Note that some factors in the constraints could be zero, effectively meaning that not every variable needs to be in every constraint.

Equality is allowed, using a sort of "trick" which you can verify. The following:

$$-a_0 + x_1 \leq 0$$

$$a_0 - x_1 \leq 0$$

is equivalent to:

$$x_1 = a_0$$

OK, so given the objective *and* the constraints, the goal of linear programming is to minimize the objective.

5.1.1 Shell

Consider a Python integer programming library where Python package known pulp can be described as:

```
$ pip install pulp
$ python
>>> from pulp import *
>>> x1 = LpVariable('x1')
>>> p = LpProblem('p', LpMinimize)
>>> p += x1
>>> p.solve()
1
>>> LpStatus
{0: 'Not Solved', 1: 'Optimal', -2: 'Unbounded', -1: 'Infeasible', -3: 'Undefined'}
>>> x1.value()
0.0
```

From the above lines:

```
$ pip install pulp
```

This simply installs pulp. If you're not in a virtualenv, you might need to use sudo.

```
>>> from pulp import *
```

This imports Pulp. I'm going to be lazy and import everything into the global namespace using *. Everything in Pulp starts with "Lp" or "lp" so it should be clear what came from the module.

```
>>> x1 = LpVariable('x1')
```

This creates a variable. The argument is the name, which is arbitrary (just make it unique if you create more than one variable).

```
>>> p = LpProblem('p', LpMinimize)
```

This creates a *problem,* which is Pulp's way of keeping pieces of the linear program together. The first argument is another arbitrary name. The second is LpMinimize because the default is to maximize the objective. (Minimizing and maximizing are mathematically equivalent — multiply the objective by -1 — so linear programming can do both).

We'll "add" the objective function with the following syntax (Pulp uses a lot of operator overloading):

```
>>> p += x1
```

Hence our objective function is as follows:

$$\text{minimize}\, obj\left(x_1\right) = x_1$$

Not very exciting, but let's see what happens.

```
>>> p.solve()
1
>>> LpStatus
{0: 'Not Solved', 1: 'Optimal', -2: 'Unbounded', -1: 'Infeasible', -3: 'Undefined'}
```

The solve function returns a status. The different statuses are described in LpStatus. In this case, Pulp is telling us it found an optimal solution to the problem. After calling solve, the variables will have their value set to whatever minimizes the objective. So let's inspect:

```
>>> x1.value()
0.0
```

Pulp tells us that the assigning 0 to x1 minimizes the objective (remember that all the variables are non-negative). In this case the problem was trivial.

So let's add a constraint:

```
>>> p += x1 >= 1
>>> p.solve()
1
>>> x1.value()
1.0
```

This adds a constraint that x1 >= 1, and then re-solves. Obviously the minimal value of x1 is now 1, since 0 is not allowed by the constraint. Now our complete linear program is:

$$\begin{aligned}&\text{minimize}\, obj\left(x_1\right) = x_1\\&\text{subject to } x_1 \geq 1\end{aligned}$$

5.2 Driver Scheduling Problem

Let's move on to integer programming, and how it gets us closer to solving our driver scheduling problem. An integer program is exactly the same as a linear program, with the additional constraint that all x are integersm *i.e,*

$$\begin{aligned}\text{minimize}\, obj\left(x_{1,}, x_2, \ldots x_n\right) = a_1x_1 + a_2x_2 + \cdots a_nx_n\\ \text{each } x_i \in \{0,1,2,\ldots\} \text{ and subject to}\\ b_0 + b_1x_1 + b_2x_2 + \cdots b_nx_n \leq 0\\ c_0 + c_1x_1 + c_2x_2 + \cdots c_nx_n \leq 0\end{aligned}$$

Now solving integer programs is an NP-hard problem where there are algorithms that can often solve them much faster than brute force. Pulp can also solve integer programs. But first, let's discuss the high level approach to how integer programs will solve our original driver scheduling problem. We'll solve our problem by modeling it as a *weighted set partitioning problem*. This image gives an intuitive idea of the problem:

(a) 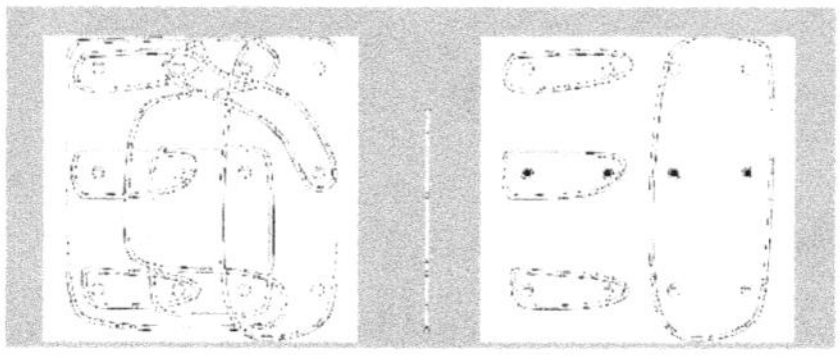 (b)

Figure 5.1: (a) Input, (b) Output.

On the left we have 12 items (the dots) and 7 sets containing different combinations of items. The solution to the set partitioning problem is the fewest number of sets which include all the items with no overlap.

The idea is that we have a set of items, represented by the points in the lefthand image above. We also have a set of sets, each represented by a squiggly loop in the Figure 5.1(a). In this case, for example, there are seven sets (trust me, since counting them is probably bad for your eyes). The set partitioning problem is to find the smallest number of sets that contain all the items (points), without any overlap between sets (hence a "partition"). Every point must be in one and exactly one set. The solution to the problem above is shown in the Figure 5.1(b).

The *weighted* set partitioning problem is when we assign a cost to each set. Then, instead of minimizing the number of sets, we minimize the sum of the values of the sets we choose. You can see that if we set all the weights to one, we get the original set partitioning problem.

Set partitioning is NP-complete (both weighted and unweighted). However, we can use integer programming to solve it optimally. NP-complete doesn't mean that problems are always slow to solve. It only means that we can construct degenerates cases that cause any general algorithm to run slowly some of the time. At least, to the best of our knowledge (unless someone proves P=NP). Integer programs can often be solved relatively quickly.

In the problem, the dots in the set partitioning diagram are trips a bus takes from its origin to its destination (our tuples), and the sets are potential driver duties. In other words, each duty is a set of trips that a driver could potentially do. Assigning weight to those sets using the driver pay rules we defined above (overtime, guarantee pay, *etc.*). When we solve the weighted set partitioning problem, we'll get the set of duties that cover all the trips and minimize the sum of individual drivers' costs. This is exactly what we want.

So here's how we'll do it. Each potential duty will be a variable (an x) in our integer program. We'll further constrain all the x to be between zero and one. Since they're integers, this actually restricts them to be *exactly* zero or one. The idea behind this is that a particular x will be *one* in the solution if the solver decides that using the duty is part of the optimal solution, and *zero* if the duty is discarded as not being part of the optimal solution.

The coefficient of each x is the cost of that duty in terms of driver pay. Hence our integer program looks like this (let's use p for pay instead of the a's we were using above):

$$\text{minimize } obj(x_1, x_2, \ldots, x_n) = p_1x_1 + p_2x_2 + \cdots p_nx_n \text{ each } x_i \in \{0,1\}$$

This means we have n potential duties. Each p is the driver pay (hence payroll cost) of the duty. We want Pulp to choose some x's to be zero and some to be one. By doing this, our objective function will turn into the sum of the payments we have to make. And we want to minimize this.

Of course, as is, there are no constraints that some x need to be selected (set to one) at all. Hence, Pulp will helpfully set all x to zero to minimize the objective.

Well, every trip needs to be in exactly one duty. So we can add a constraint for every trip, so that the x's corresponding to the duties that trip appears in sum up to 1. For example, we might have a constraint like this:

$$x_1 + x_2 + x_3 + x_4 + x_5 = 1$$

This assumes for simplicity's sake that the first five x's correspond to the duties containing trip ('A', 9). Likewise, we'd have a constraint like this for every other trip.

These constraints say that for a given trip, exactly one duty must include it. No more, no less.

5.2.1 Writing the Code

Generating every possible duty explodes exponentially with the number of trips. Hence, for now, we'll randomly generate duties, and see what happens. We already have code written above to generate the set of trips. Let's write a function to randomly generate some duties from those trips. We'll make sure that we don't generate a duty with multiple trips that occur at the same time.

So let's write it like this:

- We'll randomly choose the number of trips in the duty
- Then we'll randomly choose the hours for those trips
- Then we'll randomly choose the routes for each trip

Here's the fully functioning code to generate trips and duties (extending the code above):

$ ./duty_generation.py

[[('A', 14), ('A', 18)],

[('A', 9), ('B', 11), ('B', 12), ('D', 14), ('A', 15), ('A', 16), ('D', 18)],

[('C', 17), ('B', 18)],

[('C', 18)],

[('B', 9), ('C', 11), ('A', 16), ('B', 18), ('D', 19), ('C', 20)]]

Now write a function that tells us how much each duty costs:

And here's what the output looks like when you run it:

[(1000000.0,

[('C', 9),

('A', 10),

('C', 11),

('C', 12),

('B', 13),

('A', 14),

('B', 15),

('C', 16),

('A', 17),

('C', 18),

('B', 19),

('B', 20)]),

(8, [('B', 12), ('A', 13), ('C', 17)]),

(8, [('B', 11), ('D', 13), ('D', 14), ('A', 16), ('A', 17), ('C', 19)]),

(8, [('C', 16)]),

(11.0,

[('C', 10),

('A', 11),

('C', 12),

('D', 13),

('C', 14),

('D', 15),

('A', 16),

('A', 18),

('B', 19),

('C', 20)])]

Recall the Pulp code from above, and run it and see what happens:

driver_scheduling:

MINIMIZE

12.5*x1 + 8*x10 + 8*x11 + 8*x12 + 8*x13 + 8*x14 + 8*x15 + 8*x16 + 8*x17 + 8*x18 + 8*x19 + 9.5*x2 + 8*x20 + 8*x21 + 8*x22 + 8*x23 + 8*x24 + 8*x25 + 8*x26 + 8*x27 + 8*x28 + 8*x29 + 8*x3 + 8*x30 + 8*x31 + 8*x32 + 8*x33 + 8*x34 + 8*x35 + 8*x36 + 8*x37 + 8*x38 + 8*x39 + 1000000.0*x4 + 8*x40 + 8*x41 + 8*x42 + 8*x43 + 8*x44 + 8*x45 + 8*x46 + 8*x47 + 8*x48 + 8*x49 + 11.0*x5 + 8*x50 + 8*x51 + 8*x52 + 8*x53 + 8*x6 + 8*x7 + 8*x8 + 8*x9 + 0.0

SUBJECT TO

_C1: x7 = 1

_C2: x33 = 1

_C3: x1 + x45 = 1

_C4: x14 + x2 + x5 = 1

_C5: x1 + x20 + x4 = 1

...

_C47: x25 + x4 = 1

_C48: x24 = 1

VARIABLES

0 <= x1 <= 1 Integer

...

0 <= x9 <= 1 Integer

Optimal

Cost: 239.5

[[('B', 9),

('B', 10),

('B', 11),

('D', 12),

('D', 13),

('D', 14),

('D', 16),

('B', 17),

('D', 18),

('D', 19),

('B', 20)],

[('D', 10),

('C', 11),

('B', 13),

('B', 14),

('D', 15),

('C', 16),

('A', 17),

('B', 18),

('A', 19),

('D', 20)],

[('A', 9)],

[('A', 10)],

[('A', 11)],

[('A', 12)],

[('A', 13)],

[('A', 14)],

[('A', 15)],

[('A', 16)],

[('A', 18)],

[('A', 20)],

[('B', 12)],

[('B', 15)],

[('B', 16)],

[('B', 19)],

[('C', 9)],

[('C', 10)],

[('C', 12)],

[('C', 13)],

[('C', 14)],

[('C', 15)],

[('C', 17)],

[('C', 18)],

[('C', 19)],

[('C', 20)],

[('D', 9)],

[('D', 11)],

[('D', 17)]]

Most drivers do one trip, get paid the minimum of 8 hours, and go home. Well, it's good for them, but payroll won't be happy. This is because we're only generating 5 duties, of course. Let's increase that to 100 duties. And let's also time how long it takes. Here's the result (without Pulp output this time):

Optimal

Cost: 120

[[('A', 9), ('C', 11), ('D', 13), ('A', 14), ('D', 15)],

[('A', 10), ('D', 11), ('C', 13), ('A', 16), ('B', 18), ('D', 19)],

[('C', 17)],

[('D', 14), ('C', 15), ('D', 16), ('D', 17), ('A', 19)],

[('C', 9), ('B', 10), ('D', 18), ('C', 19), ('A', 20)],

[('B', 11), ('C', 14), ('C', 16), ('A', 18), ('B', 19), ('B', 20)],

[('D', 10), ('A', 12), ('A', 13), ('B', 16), ('A', 17), ('C', 18), ('C', 20)],

[('D', 9), ('C', 10), ('D', 12), ('B', 13), ('B', 17), ('D', 20)],

[('A', 11)],

[('A', 15)],

[('B', 9)],

[('B', 12)],

[('B', 14)],

[('B', 15)],

[('C', 12)]]

real 0m1.297s

user 0m0.110s

sys 0m0.000s

The cost is almost half! But it's still not great. Let's generate 1000 duties.

Optimal

Cost: 70.0

[[('C', 9), ('C', 11), ('C', 12), ('A', 15), ('C', 18), ('A', 19)],

[('A', 10), ('A', 11), ('D', 12), ('A', 13), ('A', 16), ('D', 18), ('C', 20)],

[('A', 9),

('C', 10),

('A', 12),

('C', 13),

('B', 14),

('D', 15),

('D', 16),

('C', 17),

('B', 19)],

[('D', 10),

('B', 12),

('B', 13),

('A', 14),

('B', 16),

('A', 17),

('B', 18),

('C', 19),

('B', 20)],

[('B', 9),

('B', 10),

('B', 11),

('D', 13),

```
('D', 14),
('B', 15),
('C', 16),
('B', 17),
('A', 18),
('D', 19)],
[('D', 9), ('D', 11), ('C', 14), ('D', 17), ('A', 20)],
[('D', 20)],
[('C', 15)]]
real 4m41.586s
user 3m11.270s
sys 0m0.990s
```

Well, we are solving an NP-complete problem after all. In practice, we could speed this up by generating duties more cleverly. For example, there's no reason for drivers to jump back and forth between routes all the time. If we limited that, we could generate more "good" duties, while ignoring "bad" duties without passing them to the solver.

5.2.2 Integer Programming as a Flexible Framework

We've used integer programming to solve our problem. And we've saved the effort of thinking up an algorithm for this specific problem, which would probably be either slow, or suboptimal, or both. Integer programming also gives us a very flexible framework for making changes, *e.g.*, if we can customize the cost function to penalize some duties (like those that switch routes too often because that's confusing). When the problem gets more complicated, such as routes ending in different locations, or trips starting and ending at different times, integer programming can handle that really easily. If you don't want something to happen you can either just add a severe penalty (like we did with the driver break rule), or add a constraint.

There are many other interesting classes of problems that can be solved using integer programming as well. Since solving an integer program is NP-hard, it can, in fact, be used to solve many NP and NP-complete problems (as long as you can model it as an integer program without exponential growth). This is a very powerful tool, because it's often much easier and effective to model an NP or NP-complete problem as an integer program than it is to write an algorithm to solve it. You can also solve polynomial time algorithms using integer programming, and this is still often useful as a first cut solution, and it often turns out to be fast enough.

To name just a few other examples of problems that integer programming is well suited to solve:

- Transportation problems (such as package delivery, or resource delivery such as energy distribution)
- Warehouse and container packing problems
- Matching problems (such as what stock brokerages do to match buyers with sellers, or scheduling which sports teams play each other)
- Map coloring

5.3 Modeling with Integer Variables

The use of integer variables in production when only integral quantities can be produced is the most obvious use of integer programs. In this section, we will look at some less obvious ones.

5.3.1 Capital Budgeting

Suppose we wish to invest \$14,000. We have identified four investment opportunities. Investment 1 requires an investment of \$5,000 and has a present value (a time-discounted value) of \$8,000; investment 2 requires \$7,000 and has a value of \$11,000; investment 3 requires \$4,000 and has a value of \$6,000; and investment 4 requires \$3,000 and has a value of \$4,000. Into which investments should we place our money so as to maximize our total present value?

As in linear programming, our first step is to decide on our variables. This can be much more difficult in integer programming because there are very clever ways to use integrality restrictions. In this case, we will use a 0-1 variable x_j for each investment. If x_j is 1 then we will make investment j. If it is 0, we will not make the investment. This leads to the 0-1 programming problem:

$$\text{Maximize } 8x_1 + 11x_2 + 6x_3 + 4x_4$$
$$\text{subject to } 5x_1 + 7x_2 + 4x_3 + 3x_4 \leq 14$$
$$x_j \in \{0,1\}\ j = 1, \ldots 4.$$

Now, a straightforward ``bang for buck'' suggests that investment 1 is the best choice. In fact, ignoring integrality constraints, the optimal linear programming solution is $x_1 = 1$, $x_2 = 1$, $x_3 = 0.5$, $x_4 = 0$ for a value of \$22,000. Unfortunately, this solution is not integral. Rounding x_3 down to 0 gives a feasible solution with a value of \$19,000. There is a better integer solution, however, of $x_1 = 0$, $x_2 = x_3 = x_4 = 1$ for a value of \$21,000. This example shows that rounding does not necessarily give an optimal value.

There are a number of additional constraints we might want to add. For instance, consider the following constraints:

1. We can only make two investments.
2. If investment 2 is made, investment 4 must also be made.
3. If investment 1 is made, investment 3 cannot be made.

All of these, and many more *logical restrictions,* can be enforced using 0-1 variables. In these cases, the constraints are:

1. $x_1 + x_2 + x_3 + x_4 \le 2$
2. $x_2 - x_4 \le 0$
3. $x_1 + x_3 \le 1.$

Exercise 1 (Optional) As the leader of an oil exploration drilling venture, you must determine the best selection of 5 out to 10 possible sites. label the sites $s_1, s_2, \ldots, s_{10}$ and the expected profits associated with each as $p_1, p_2, \ldots, p_{10}$.

i) If site s_2 is explored, then site s_3 must also be explored. Furthermor, regional development restrictions are such that
ii) Exploring sites s_1 and s_7 will prevent you from exploring site s_8.
iii) Exploring sites s_3 or s_4 will prevent you from exploring site s_5.

Formulate an integer program to determine the best exploration scheme.

In the preceding, we considered making a single investment in a project for the duration of some term, and receiving its return at the end of the term. In practice, we may face a choice among projects that require investments of different amounts in each of several periods (with possibly different budgets available in each period), with the return being realized over the life of the project. In this case, we can still model the problem with variables

$$x_j = \begin{cases} 1 & \text{if we invest in project } j \\ 0 & \text{otherwise,} \end{cases}$$

the objective is still to maximize the sum of the returns on the projects selected, and there is now a budget constraint for each period. For example, suppose we wish to invest \$14,000, \$12,000, and \$15,000 in each month of the next quarter. We have identified four investment opportunities. Investment 1 requires an investment of \$5,000, \$8,000, and \$2,000 in month 1, 2, and 3, respectivey, and has a present value (a time-discounted value) of \$8,000; investment 2 requires \$7,000 in month 1 and \$10,000 in period 3, and has a value of \$11,000; investment 3 requires \$4,000 in period 2 and \$6,000 in period 3, and has a value of \$6,000; and investment 4 requires \$3,000, \$ 4,000, and \$5,000 and has a value of \$4,000. The corresponding integer program is

$$\begin{aligned} \text{Maximize} \quad & 8x_1 + 11x_2 + 6x_3 + 4x_4 \\ \text{subject to} \quad & 5x_1 + 7x_2 \qquad\quad + 3x_4 \le 14 \\ & 8x_1 \qquad\quad + 4x_3 + 4x_4 \le 12 \\ & 2x_1 + 10x_2 + 6x_3 + 4x_4 \le 15 \\ & x_j \in \{0, 1\}. \end{aligned}$$

5.3.2 Knapsack

The knapsack problem is a particularly simple integer program: it has only one constraint. Furthermore, the coefficients of this constraint and the objective are all non-negative. For example, the following is a knapsack problem:

$$\begin{array}{ll} \text{Maximize} & 8x_1 + 11x_2 + 6x_3 + 4x_4 \\ \text{subject to} & 5x_1 + 7x_2 + 4x_3 + 3x_4 \leq 14 \\ & x_j \in \{0, 1\}. \end{array}$$

The traditional story is that there is a knapsack (here of capacity 14). There are a number of items (here there are four items), each with a size and a value (here item 2 has size 7 and value 11). The objective is to maximize the total value of the items in the knapsack. Knapsack problems are nice because they are (usually) easy to solve, as we will see in the dynamic programming section of this course.

To solve the associated linear program, it is simply a matter of determining which variable gives the most ``bang for the buck''. If you take C_j/a_j (the objective coefficient/constraint coefficient) for each variable, the one with the highest ratio is the best item to place in the knapsack. Then the item with the second highest ratio is put in and so on until we reach an item that cannot fit. At this point, a fractional amount of that item is placed in the knapsack to completely fill it. In our example, the variables are already ordered by the ratio. We would first place item 1 in, then item 2, but item 3 doesn't fit: there are only 2 units left, but item 3 has size 4. Therefore, we take half of item 3. The solution $x_1 = 1$ $x_2 = 1$, $x_3 = 0.5$, $x_4 = 0$ is the optimal solution to the linear program.

5.3.3 Lockbox Problem

Consider a national firm that receives checks from all over the United States. Due to the vagaries of the U.S. Postal Service, as well as the banking system, there is a variable delay from when the check is postmarked (and hence the customer has met her obligation) and when the check clears (and when the firm can use the money). For instance, a check mailed in Pittsburgh sent to a Pittsburgh address might clear in just two days. A similar check sent to Los Angeles might take eight days to clear. It is in the firm's interest to have the check clear as quickly as possible since then the firm can use the money. In order to speed up this clearing, firms open offices (called lockboxes) in different cities to handle the checks.

For example, suppose we receive payments from four regions (West, Midwest, East, and South). The average daily value from each region is as follows: \$70,000 from the West, \$50,000 from the Midwest, \$60,000 from the East, and \$40,000 from the South. We are considering opening lockboxes in L.A., Chicago, New York, and/or Atlanta. Operating a lockbox costs \$50,000 per year. The average days from mailing to clearing is given in the table. Which lockboxes should we open?

Table 5.1: Clearing Times

From	*L.A.*	*Chicago*	*New York*	*Atlanta*
West	2	6	8	8
Midwest	6	2	5	5
East	8	5	2	5
South	8	5	5	2

First we must calculate the losses due to lost interest for each possible assignment. For example, if the West sends to New York, then on average there will be $560,000 (= 8 × $70,000) in process on any given day. Assuming an investment rate of 20%, this corresponds to a yearly loss of $112,000. We can calculate the losses for the other possibilities in a similar fashion to get table below.

Table 5.2: Lost Interest ($1000)

From	*L.A.*	*Chicago*	*New York*	*Atlanta*
West	28	84	112	112
Midwest	60	20	50	50
East	96	60	24	60
South	64	40	40	16

The formulation takes a bit of thought. Let y_j be a 0-1 variable that is 1 if lockbox j is opened and 0 if it is not. Let x_{ij} be 1 if region i sends to lockbox j.

Our objective is to minimize our total yearly costs. This is:

$$28x_{11} + 84x_{12} + 112x_{13} + 112x_{14} + 60x_{21} + \ldots + 50y_1 + 50y_2 + 50y_3 + 50y_4$$

One set of constraints is as follows:

$$\sum_j x_{ij} = 1 \text{ for all } i$$

(each region must be assigned to one lockbox).

A more difficult set of constraints is that a region can only be assigned to an open lockbox. For lockbox 1 (L.A.), this can be written

$$x_{11} + x_{21} + x_{31} + x_{41} \leq 100y_1$$

(There is nothing special about 100; any number at least 4 would do.) Suppose we do not open L.A. Then y_1 is 0, so all of x_{11}, x_{21}, x_{31}, and x_{41} must also be. If y_1 is 1 then there is no restriction on the x values.

We can create constraints for the other lockboxes to finish off the integer program. For this problem, we would have twenty variables (four y variables, sixteen x variables) and eight constraints. This gives the following 0-1 linear program:

$$
\begin{aligned}
\text{Minimize} \quad & 28x_{11} + 84x_{12} + 12x_{13} + 112x_{14} + 60x_{21} + 20x_{22} + 50x_{23} + 50x_{24} \\
& + 96x_{31} + 60x_{32} + 24x_{33} + 60x_{34} + 60x_{41} + 40x_{42} + 40x_{43} + 16x_{44} \\
& + 50y_1 + 50y_2 + 50y_3 + 50y_4 \\
\text{subject to} \quad & x_{11} + x_{12} + x_{13} + x_{14} = 1 \\
& x_{21} + x_{22} + x_{23} + x_{24} = 1 \\
& x_{31} + x_{32} + x_{33} + x_{34} = 1 \\
& x_{41} + x_{42} + x_{43} + x_{44} = 1 \\
& x_{11} + x_{21} + x_{31} + x_{41} = 1 \leq 100y_1 \\
& x_{12} + x_{22} + x_{32} + x_{42} = 1 \leq 100y_2 \\
& x_{13} + x_{23} + x_{33} + x_{43} = 1 \leq 100y_3 \\
& x_{14} + x_{24} + x_{34} + x_{44} = 1 \leq 100y_4 \\
& x_{ij}, y_i \in \{0.1\} i = 1, \ldots, 4, j = 1, \ldots, 4
\end{aligned}
$$

A possible LINGO model is

```
model:
sets:
Region /West, Midwest, East, South/;
Box /LA, Chi, NY, Atl/: open, cost;
Link (Region, Box): assign, loss;
endsets
min = @sum(Link: loss * assign) + @sum(Box: cost * open);
@for (Region(i):
@sum(Box(j): assign(i,j)) = 1
);
@for (Box(j):
@sum(Region(i): assign(i,j)) < big * open(j);
);
@for(Link: @bin(assign));
@for(Box: @bin(open));
data:
big = 100;
cost = 50 50 50 50;
loss = 28 84 112 112
60 20 50 50
96 60 24 60
64 40 40 16;
enddata
end
```

If we ignore integrality, we get the solution $x_{11} = x_{22} = x_{33} = x_{44} = 1$, $y_1 = y_2 = y_3 = y_4 = 0.01$ and the rest equal to 0. Note that we get no useful information out of this linear programming solution.

The above is a perfectly reasonable 0-1 programming formulation of the lockbox problem. Note that many variations are possible (New York costs more to operate in than other cities, South won't send to L.A., and so on).

There are other formulations, however. For instance, consider the sixteen constraints of the form

$$x_{ij} \leq y_i$$

These constraints also force a region to only use open lockboxes (check this!). It might seem that a larger formulation is less efficient and therefore should be avoided. This is not the case! If we solve the linear program with the above constraints, we get the solution $x_{11} = x_{22} = x_{33} = x_{44} = 1$, $y_1 = y_4 = 1$ with the rest equal to zero. In fact, we have an integer solution, which must therefore be optimal! Different formulations can have very different properties with respect to their associated linear program. One very active research area is to take common problems and find good reformulations.

The above is an example of a *fixed charge* problem. There is a fixed charge for opening a lockbox, but then it can be used as much as desired. There are many other types of fixed charge problems, including the plant location problems described in Winston. Similar techniques also apply for a variety of constraints, such as batch size constraints (where if an activity is engaged in at all, it must have a minimum level, *i.e.*, $x_j = 0$ or $x_j \geq l_j$) and other types of "either-or" or *disjunctive* constraints.

5.3.4 Set Covering

To illustrate this model, consider the following *location* problem: A city is reviewing the location of its fire stations. The city is made up of a number of neighborhoods, as illustrated in Figure 5.2.

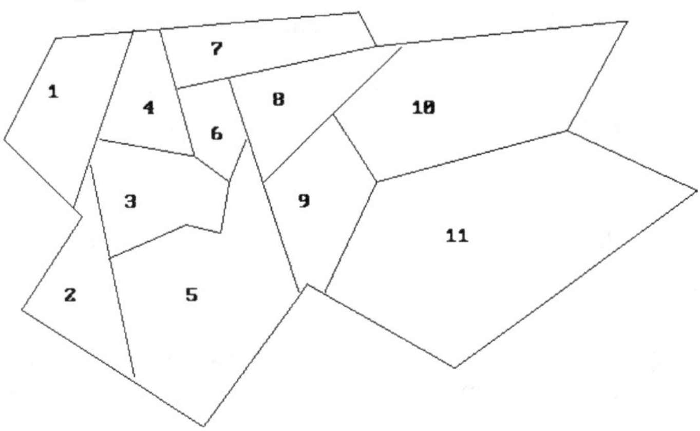

Figure 5.2: Map of the City.

A fire station can be placed in any neighborhood. It is able to handle the fires for both its neighborhood and any adjacent neighborhood (any neighborhood with a non-zero border with its home neighborhood). The objective is to minimize the number of fire stations used.

We can create one variable x_j for each neighborhood j. This variable will be 1 if we place a station in the neighborhood, and will be 0 otherwise. This leads to the following formulation

$$
\begin{array}{ll}
\text{Minimize} & x_1 + x_2 + x_3 + x_4 + x_5 + x_6 + x_7 + x_8 + x_9 + x_{10} + x_{11} \\
\text{subject to} & x_1 + x_2 + x_3 + x_4 \geq 1 \\
& x_1 + x_2 + x_3 + x_5 \geq 1 \\
& x_1 + x_2 + x_3 + x_4 + x_5 + x_6 \geq 1 \\
& x_1 + x_3 + x_4 + x_6 + x_7 \geq 1 \\
& x_2 + x_3 \quad x_5 + x_6 + x_8 + x_9 \geq 1 \\
& + x_3 + x_4 + x_5 + x_6 + x_7 + x_8 \geq 1 \\
& + x_4 + x_6 + x_7 + x_8 \geq 1 \\
& + x_5 + x_6 + x_7 + x_8 + x_9 + x_{10} \geq 1 \\
& + x_5 + x_8 + x_9 + x_{10} + x_{11} \geq 1 \\
& x_8 + x_9 + x_{10} + x_{11} \geq 1 \\
& x_9 + x_{10} + x_{11} \geq 1
\end{array}
$$

$$x_j \in \{0, 1\} \; j = 1, \ldots, 11$$

The first constraint states that there must be a station either in neighborhood 1 or in some adjacent neighborhood. The next constraint is for neighborhood 2 and so on. Notice that the constraint coefficient a_{ij} is 1 if neighborhood i is adjacent to neighborhood j or if $i=j$ and 0 otherwise. The jth column of the constraint matrix represents the set of neighborhoods that can be served by a fire station in neighborhood j. We are asked to find a set of such subsets j that *covers* the set of all neighborhoods in the sense that every neighborhood appears in the service subset associated with *at least* one fire station.

One optimal solution to this is $x_3 = x_8 = x_9 = 1$ and the rest equal to 0.

This is an example of the *set covering problem*. The set covering problem is characterized by having binary variables, ≥ constraints each with a right hand side of 1, and having simply sums of variables as constraints. In general, the objective function can have any coefficients, though here it is of a particularly simple form.

5.3.5 Set Packing and Partitioning

Other problems involving relationships between a set and some of its subsets are related to set covering. The *set packing problem* arises when each set element must appear in *at most* one subset. In this case, the constraints are of the less-than-or-equal form. The *set partitioning problem* arises when each set element must appear in *exactly* one subset, and the constraints in this problem are equality constraints.

One interesting area where set partitioning is used is in airline crew scheduling. Next to fuel costs, personnel costs are the largest cost an airline faces; captain salaries of $150,000 per year are not unusual. The effective use of personnel can have a tremendous impact on the bottom line. There are, however, a tremendous number of restrictions on how airline personnel are used. Between FAA regulations and union contracts, determining a feasible crew schedule is a very difficult problem. About ten years ago, American Airline was interested in applying OR techniques to crew scheduling. In their formulation, a variable is created for each feasible schedule for a single crew. For instance, a schedule where the crew leaves Pittsburgh at 7:15AM on Tuesday, arrives in NY at 8:30, leaves NY at 9:00, arrives Atlanta at 11:00, leaves Atlanta at 12:00 and arrives Pittsburgh at 2:30 would correspond to a single variable, x_1, say. The cost of setting that variable to 1 (meaning to have a crew fly that schedule) would be based on the union contract and other regulations.

As you can imagine, there are a tremendous number of variables. The constraints, however are very simple. For each flight (called a *leg*), it is necessary to choose a schedule that includes that flight. This leads to a constraint of the form

$$x_1 + x_4 + x_{55} + x_{5534} + \dots = 1$$

if schedules 1, 4, 55, 5534 and so on all contain the 7:15 Tuesday flight from Pittsburgh to NY (for example). While the resulting is very large, the savings that result from finding good solutions are so large that a tremendous amount of work has gone into finding techniques for solving these problems.

5.4 Features about Traveling Salesperson Problem

Consider a traveling salesperson who must visit each of 20 cities before returning home. She knows the distance between each of the cities and wishes to minimize the total distance traveled while visiting all of the cities. In what order should she visit the cities?

Let there be n cities, numbered from 1 up to n. For each pair of cities (i,j) let c_{ij} be the cost of going from city i to city j or from city j to city i. Let's let x_{ij} be 1 if the person travels between cities i and j (either from city i to city j or from j to i). This problem is known as the *symmetric* TSP. In the *asymmetric* TSP, the cost to travel in one direction may differ from the cost to travel in the other, and the decision variables must distinguish between the two directions. Clearly the asymetric problem is the more general.

The objective is to minimize $\sum_{i=1}^{n}\sum_{j=1}^{i-1} c_{ij} x_{ij}$. The constraints are harder to find. Consider the following set:

$$\sum_{j \neq i} x_{ij} = 2 \text{ for all } i$$

These constraints say that every city must be visited. These constraints are not enough, however, since it is possible to have multiple cycles (*subtours*), rather than one big cycle (*tour*) through all the points. To handle this condition, we can use the following set of *subtour elimination constraints*:

$$\sum_{i \in S}\sum_{j \notin S} x_{ij} \geq 2 \text{ for all } S \subset N.$$

This set states that, for any subset of cities S, the tour must enter and exit that set. These, together with $x_{ij} \in \{0,1\}$ is sufficient to formulate the traveling salesperson problem (TSP) as an integer program.

Note, however, that there are a tremendous number of constraints: for our 20-city problem, that number is roughly 524,288. For a 300-city problem, this would amount to 101851798816724304313422284420468908052573419683296

812531807022467719064988166835309169868 8 constraints. Try putting that into LINGO!

Despite the apparent complexity of this formulation, it lies at the heart of the most promising current approach to solving medium (100-1000 city) TSPs. We will see this again in the cutting plane section. Other forms of subtour elimination constraints are possible, but they suffer from the same explosive rate of growth in number. Subtour elimination constraints for the asymetric problem are equally difficult.

5.5 Mixed-Integer Programming

The problems most commonly solved by the Gurobi Parallel Mixed Integer Programming solver are of the form:

Objective: minimize $c^T x$

Constraints: $A x = b$ (linear constraints)

$l \leq x \leq u$ (bound constraints)

some or all xj must take integer values (integrality constraints)

The integrality constraints allow MIP models to capture the discrete nature of some decisions. For example, a variable whose values are restricted to 0 or 1, called a binary variable, can be used to decide whether or not some action is taken, such as building a warehouse or purchasing a new machine.

The Gurobi MIP solver can also solve models with a quadratic objective and/or quadratic constraints:

Objective: minimize $x^T Q x + q^T x$
Constraints: $A x = b$ (linear constraints)
$l \leq x \leq u$ (bound constraints)
$x^T Q_i x + q_i^T x \leq b_i$ (quadratic constraints)
some or all x must take integer values (integrality constraints)

MIP models with a quadratic objective but without quadratic constraints are called Mixed Integer Quadratic Programming (MIQP) problems. MIP models with quadratic constraints are called Mixed Integer Quadratically Constrained Programming (MIQCP) problems. Models without any quadratic features are often referred to as Mixed Integer Linear Programming (MILP) problems.

What follows is a description of the algorithm used by Gurobi to solve MILP models. The extension to MIQP and MIQCP is mostly straightforward, but we won't describe them here.

5.5.1 Branch-and-Bound

Mixed Integer Linear Programming problems are generally solved using a linear-programming based branch-and-bound algorithm. Basic LP-based branch-and-bound can be described as follows. We begin with the original MIP. Not knowing how to solve this problem directly, we remove all of the integrality restrictions. The resulting LP is called the linear-programming *relaxation* of the original MIP. We can then solve this LP. If the result happens to satisfy all of the integrality restrictions, even though these were not explicitly imposed, then we have been quite lucky. This solution is an optimal solution of the original MIP, and we can stop. If not, as is usually the case, then the normal procedure is to pick some variable that is restricted to be integer, but whose value in the LP relaxation is fractional. For the sake of argument, suppose that this variable is x and its value in the LP relaxation is 5.7. We can then exclude this value by, in turn, imposing the restrictions $x \leq 5.0$ and $x \geq 6.0$.

If the original MIP is denoted P_0, then we might denote these two new MIPs by P_1, where $x \leq 5.0$ is imposed, and P_2, where $x \geq 6.0$ is imposed. The variable x is then called a *branching variable*, and we are said to have *branched* on x, producing the two sub-MIPs P_1 and P_2. It should be clear that if we can compute optimal solutions for each of P_1 and P_2, then we can take the better of these two solutions and it will be optimal to the original problem, P_0. In this way we have replaced P_0 by two simpler (or at least more-restricted) MIPs. We now apply the same idea to these two MIPs, solving the corresponding LP relaxations and, if necessary, selecting branching variables. In so doing we generate what is called a *search tree*. The MIPs generated by the search procedure are called the *nodes* of the tree, with P_0 designated as the *root node*. The leaves of the tree are all the nodes from which we have not yet branched. In general, if we reach a point at which we can solve or otherwise dispose of all leaf nodes, then we will have solved the original MIP.

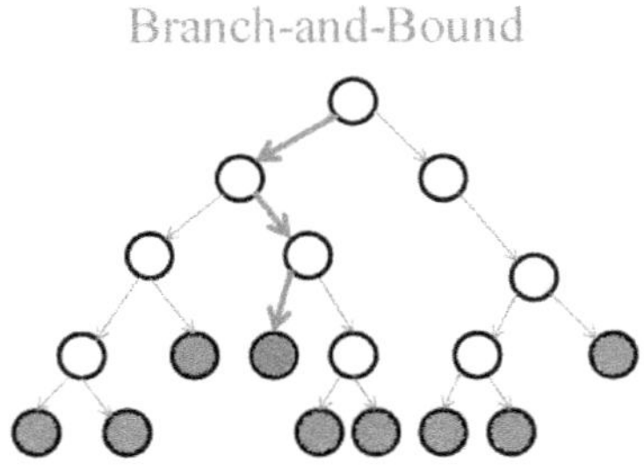

Figure 5.3: Each Node in Branch-and-Bound is a New MIP.

5.5.2 Fathomed and Incumbent Nodes

To complete our description of (LP-based) branch-and-bound we need to describe the additional logic that is applied in processing the nodes of the search tree. Let us assume that our goal is to minimize the objective, and suppose that we have just solved the LP relaxation of some node in the search tree. If it happens that all of the integrality restrictions in the original MIP are satisfied in the solution at this node, then we know we have found a feasible solution to the original MIP. There are two important steps that we then take.

First, we designate this node as *fathomed*. It is not necessary to branch on this node; it is a permanent leaf of the search tree. Second, we analyze the information provided by the feasible solution we have just found, as follows. Let us denote the best integer solution found at any point in the search as the *incumbent*. At the start of the search, we have no incumbent. If the integer feasible solution that we have just found has a better objective function value than the current incumbent (or if we have no incumbent), then we record this solution as the new incumbent, along with its objective function value. Otherwise, no incumbent update is necessary and we simply proceed with the search.

There are two other possibilities that can lead to a node being fathomed. First, it can happen that the branch that led to the current node added a restriction that made the LP relaxation infeasible. Obviously if this node contains no feasible solution to the LP relaxation, then it contains no integer feasible solution. The second possibility is that an optimal relaxation solution is found, but its objective value is bigger than that of the current incumbent. Clearly this node cannot yield a better integral solution and again can be fathomed.

5.5.3 Best Bound and Gap

There are two additional important values we need to introduce to complete our description of branch-and-bound. First observe that, once we have an incumbent, the objective value for this incumbent, assuming the original MIP is a minimization problem, is a valid upper bound on the optimal solution of the given MIP. That is, we know that we will never have to accept an integer solution of value higher than this value. Somewhat less obvious is that, at any time during the branch-and-bound search we also have a valid lower bound, sometimes call the *best bound*. This bound is obtained by taking the minimum of the optimal objective values of all of the current leaf nodes. Finally, the difference between the current upper and lower bounds is known as the *gap*. When the gap is zero we have demonstrated optimality.

5.5.4 Presolve, Cutting Planes, Heuristics, Parallelism

The field of mixed integer programming has witnessed remarkable improvements in recent years in the capabilities of MIP algorithms. Four of the biggest contributors have been *presolve, cutting planes, heuristics, and parallelism*. We now give high-level overviews of these four components.

5.5.4.1 Presolve

Presolve refers to a collection of problem reductions that are typically applied in advance of the start of the branch-and-bound procedure. These reductions are intended to reduce the size of the problem and to tighten its formulation.

A simple example of a size-reducing transformation is given below:

Suppose a given problem contains the following constraints:

$x_1 + x_2 + x_3 \geq 15$

$x_1 \leq 7$

$x_2 \leq 3$

$x_3 \leq 5$

Clearly the only way that all of these constraints can be satisfied is if $x_1 = 7$, $x_2 = 3$, and $x_3 =5$. In this case we can substitute out these variables, completely removing them from the formulation along with the above four constraints. The list of such possible reductions, of which this is only one, is quite extensive and can have an enormous effect on the overall size of the problem.

The above reduction is what we would call an LP-presolve reduction, since its validity does not depend on integrality restrictions. An example of an MIP-specific reduction is the following. Suppose that x_1 and x_2 are non-negative integer variables and that our formulation includes a constraint of the following form:

$2\,x_1 + 2\,x_2 \leq 1.$

Dividing both sides of this constraint by 2 yields:

$x_1 + x_2 \leq ½.$

Since x_1 and x_2 are both required to be integral, this inequality clearly implies that $x_1 + x_2 \leq 0$, and so by non-negativity that $x_1 = x_2 = 0$. Hence both of these variables and this constraint can be removed from the formulation. Note also that this reduction is different in character from the first in the sense that we have actually reduced the set of feasible solutions to the LP relaxation, even though the set of integer feasible solutions has remained the same. This kind of tightening can be critical to the solution of an integer program, and is one of the reasons that MIP presolve is an important tool in the solution on MIPs, much more so than LP presolve in the solution of linear programs.

5.5.4.2 Cutting Planes

Let us now consider the idea of cuttings planes. We should say at the outset that the theory of cutting planes is deep and extensive. It is also generally accepted to be the single most important contributor to the computational advances that have been made in integer programming over the last several years. The idea of cutting planes is that they tighten the formulation by removing undesirable fractional solutions, as in the case of MIP presolve, but that they do this during the solution process and without the undesirable side-effect of creating additional sub-problems (unlike branching).

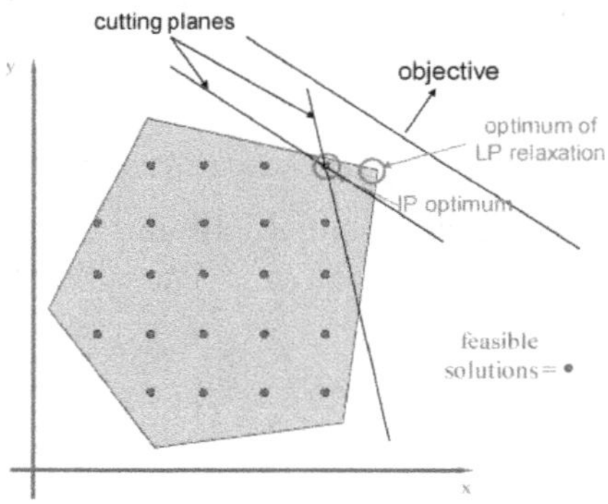

Figure 5.4: Cutting Planes.

Here is one simple example of a cutting plane. Suppose our formulation includes the following constraint:

$$6\,x_1 + 5\,x_2 + 7\,x_3 + 4\,x_4 + 5\,x_5 \leq 15,$$

where x_1 through x_5 are restricted to be binary. Suppose in addition that we have just solved an LP relaxation and that these variables take the following values in this LP relaxation: $x_1 = 0$, $x_2 = 1$, $x_3 = x_4 = x_5 = 3/4$. This undesirable solution can be excluded with the following observation: since $7 + 4 + 5 = 16 > 15$, it is not possible that $x_3 = x_4 = x_5 = 1$, and hence that the following new inequality is a valid addition to the given MIP: $x_3 + x_4 + x_5 \leq 2$. Since $3/4 + 3/4 + 3/4 = 9/4 > 2$, the new inequality cuts off the current solution. This inequality is an example of a so-called *knapsack cover*.

The reader may ask at this point why we have not simply added this new constraint, or cut, at the start. There are several reasons. One is that there are generally an enormous number of such additional constraints. It would be too expensive to find them all, and likely impossible to add them all to the model. A second reason is that adding constraints makes the LP relaxations progressively harder to solve. We only want to add these constraints if we know they will help.

Judiciously adding such constraints can have an enormously beneficial effect on the solution process.

5.5.4.3 Heuristics

We introduced the concept of the incumbent in our introduction to branch-and-bound. Having good incumbents, and finding them as quickly as possible, can be extremely valuable in the MIP search for a number of reasons.

First, it may not be possible to solve a problem to provable optimality. For example, the underlying MIP may just be too difficult, or there may be some user imposed restriction on the amount of time that we can allow our MIP algorithm run. In either case, we want to have the best possible feasible solution at termination. Having good feasible solutions also helps the search process prior to termination. The better the objective value of the incumbent, the more likely it is that the value of an LP relaxation will exceed it (in a minimization problem) and hence lead to a node being fathomed.

As the above remarks are meant to suggest, it has turned out to be extremely valuable to do a little extra work at some of the nodes of the search tree to see if a good integer feasible solution can be extracted, even though integrality has not yet resulted due to the branching. For example, it may be that many of the integer variables, while not integral, have values that are quite close to integral. We could then consider rounding some of these variables to their nearby values, fixing them to these values, solving the resulting LP relaxation, and repeating this procedure several times in the hopes that all integer variables will fall into line. If they do, and if the resulting feasible has a better objective value than the current incumbent, we can replace that incumbent and proceed.

5.5.4.4 Parallelism

As noted at the beginning of this discussion, the Gurobi MIP solver runs in parallel. The main source of parallelism is the fact that different nodes in the MIP tree search can be processed independently. As is probably apparent, however, the root node presents limited parallelism opportunities. Thus, models that explore large search trees can exploit cores quite effectively, while those that spend the majority of their runtime at the root node are more constrained in their ability to utilize multiple cores.

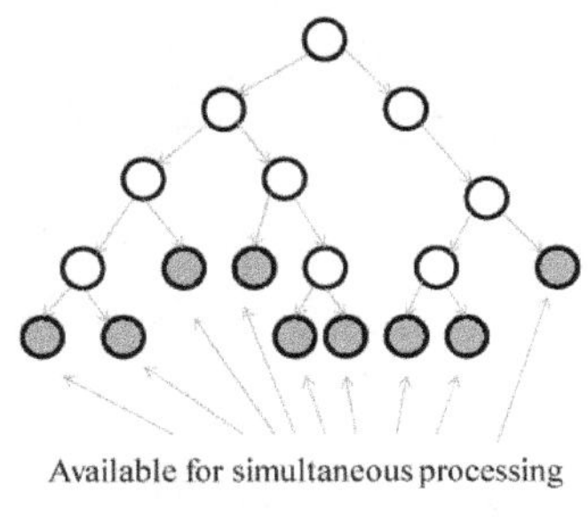

Figure 5.5: Parallel Branch and Bound.

The behaviors of most of the strategies and techniques described here can be adjusted using Gurobi parameters. While some models can benefit from parameter tuning, our goal in designing and building the Gurobi Optimizer has been to make the default settings work as well as possible across a broad range of models.

UNIT 6

Constraint Programming

6.1 Introduction

A constraint is simply a logical relation among several unknowns (or variables), each taking a value in a given domain. A constraint thus restricts the possible values that variables can take, it represents some partial information about the variables of interest. For instance, "the circle is inside the square" relates two objects without precisely specifying their positions, *i.e.*, their coordinates. Now, one may move the square or the circle and he or she is still able to maintain the relation between these two objects. Also, one may want to add other object, say triangle, and introduce another constraint, say "square is to the left of the triangle". From the user (human) point of view, everything remains absolutely transparent.

Constraints naturally enjoy several interesting properties:

- constraints may specify partial information, *i.e.*, constraint need not uniquely specify the values of its variables,
- constraints are non-directional, typically a constraint on (say) two variables X, Y can be used to infer a constraint on X given a constraint on Y and vice versa,
- constraints are declarative, *i.e.*, they specify what relationship must hold without specifying a computational procedure to enforce that relationship,
- constraints are additive, *i.e.*, the order of imposition of constraints does not matter, all that matters at the end is that the conjunction of constraints is in effect,
- constraints are rarely independent, typicaly constraints in the constraint store share variables.

Constraints arise naturally in most areas of human endeavor. The three angles of a triangle sum to 180 degrees, the sum of the currents floating into a node must equal zero, the position of the scroller in the window scrollbar must reflect the visible part of the underlying document, these are some examples of constraints which appear in the real world. Thus, constraints are a natural medium for people to express problems in many fields.

Constraint optimization, or constraint programming (CP), is the name given to identifying feasible solutions out of a very large set of candidates, where the problem can be modeled in terms of arbitrary constraints. CP problems arise in many scientific and engineering disciplines. The word "programming" is a bit of a misnomer, similar to how "computer" once meant "a person who computes". Here, "programming" refers to the arrangement of a plan, rather than programming in a computer language.

CP is based on feasibility (finding a feasible solution) rather than optimization (finding an optimal solution) and focuses on the constraints and variables rather than the objective function. In fact, a CP problem may not even have an objective function — the goal may simply be to narrow down a vary large set of possible solutions to a more manageable subset by adding constraints to the problem.

An example of a problem that is well-suited for CP is employee scheduling. The problem arises when companies that operate continuously — such as factories — need to create weekly schedules for their employees.

6.1.1 Example

A company runs three 8-hour shifts per day and assigns three of its four employees to different shifts each day, while giving the fourth the day off. Even in such a small case, the number of possible schedules is huge: each day, there are $4! = 4 \cdot 3 \cdot 2 \cdot 1 = 24$ possible employee assignments, so the number of possible weekly schedules is 247, which is over 4.5 billion. Usually there will be other constraints that reduce the number of feasible solutions — for example, that each employee works at least a minimum number of days per week. The CP method keeps track of which solutions remain feasible when you add new constraints, which makes it a powerful tool for solving large, real-world scheduling problems.

CP is a relatively new field, but has a widespread and very active community around the world with dedicated scientific journals, conferences, and an arsenal of different solving techniques. CP has been successfully applied in planning, scheduling, and numerous other domains with heterogeneous constraints. For example, the solution methods of Constraint Satisfaction Problem (CSP) can be given easily as follows:

- Various algorithms, Linear Programming, Newton method, Dynamic Programming, *etc.*, which have been researched by Operations Research (OR).
- Various techniques based on Tree Reference.
- Techniques based on constraint rewriting such as the merging method or condensing by the programming conversion.
- Algorithms for Solution Improvement such as minimum-conflict, taboo search, Simulated Annealing (SA).
- Search by using neural network.
- Genetic Algorithm (GA).

The follows can be given as Constraint Paradigm in a wide sense:

- Theory of Constraint (TOC) that attracts notice because of SCM and so on.
- Syntax theory (HPSG, Constraint Grammar *etc.*) based on constraint in Natural Programming Language Processing

It is meaningless to strictly define Constraint Programming, because there are various viewpoints toward "What is Constraint Programming or not". However, the following points can be generally given as for the center point of the area covered by Constraint Programming.

- Constraint is not limited to a numerical value. The problem is treated as a constraint irrespective of a computational area of problem.
- The central technology is the effective algorithm to keep matching by detecting a contradiction included in the problem space.
- Tree search is the base for searching solution and thus the basis is reducing a search space by positively using constraints when searching.

The especial important point seems to be the paradigm-like side, on which Constraint Programming handles the problem as a constraint regardless of the computational area of problem.

To be concrete, only the specific area out of the computational areas is focused on, researched and applied. When the superior technique is suggested in the related area, the paradigm based on constraint can be given as a framework for taking in such a technique.

In this sense, Constraint Programming in a wide sense plays a part as a node to connect the techniques that have researched in various fields. Therefore this site will take up not only Constraint Programming in a narrow sense but also paradigm for Constraint Satisfaction Problem (CSP) in a wide sense and Constraint Paradigm as possible.

Constraint Logic Programming is what the above Constraint Programming is realized as an extension of Logic Programming. Logic Programming has many features jointly with Constraint Programming and so Logic Programming can be said to be a part of Constraint Programming. Actually a lot of prolog programming systems, which are the representative of Logic Programming Languages, have constraint solvers at present. Many systems for Constraint Programming are based on the idea of Logic Programming even if they are always not based on Prolog. Constraint Programming in a narrow sense can be said to be the same as Constraint Logic Programming.

6.1.2 Constraint Satisfaction

Constraint Satisfaction arose from the research in Artificial Intelligence (combinatorial problems, search) and Computer Graphics (SKETCHPAD system, expressing geometric coherence in the case of scene analysis). The Constraint Satisfaction Problem (CSP) is a problem where one is given:

- a finite set of variables,
- a function which maps every variable to a finite domain,
- a finite set of constraints.

Each constraint restricts the combination of values that a set of variables may take simultaneously. A solution of a CSP is an assignment to each variable a value from its domain satisfying all the constraints. The task is to find one solution or all solutions.

Thus, the CSP is a combinatorial problem which can be solved by search. There exists a trivial algorithm that solves such problems or finds that there is no solution. This algorithm generates all possible combinations of values and, then, it tests whether the given combination of values satisfies all constraints or not (consequently, this algorithm is called generate and test). Clearly, this algorithm takes a long time to run so the research in the area of constraint satisfaction concentrate on finding algorithms which solve the problem more efficiently, at least for a given subset of problems.

Constraint Satisfaction Problems (CSP) have been a subject of research in Artificial Intelligence for many years. The pioneering works on networks of constraints were motivated mainly by problems arising in the field of picture processing. AI research, concentrated on difficult combinatorial problems, is dated back to sixties and seventies and it has contributed to considerable progress in constraint-based reasoning. Many powerful algorithms was designed that became a basis of current constraint satisfaction algorithms.

Constraint Satisfaction Problem (CSP) consist of:

- a set of variables X={x1,...,xn},
- for each variable xi, a finite set Di of possible values (its domain),
- and a set of constraints restricting the values that the variables can simultaneously take.

Note that values need not be a set of consecutive integers (although often they are), they need not even be numeric. A solution to a CSP is an assignment of a value from its domain to every variable, in such a way that every constraint is satisfied. We may want to find:

- just one solution, with no preference as to which one,
- all solutions,
- an optimal, or at least a good solution, given some objective function defined in terms of some or all of the variables.

Solutions to CSPs can be found by searching systematically through the possible assignments of values to variables. Search methods divide into two broad classes, those that traverse the space of partial solutions (or partial value assignments), and those that explore the space of complete value assignments (to all variables) stochastically.

6.1.3 Reasons

The reasons for choosing to represent and solve a problem as a CSP rather than, say as a mathematical programming problem are twofold.

- Firstly, the representation as a CSP is often much closer to the original problem: the variables of the CSP directly correspond to problem entities, and the constraints can be expressed without having to be translated into linear inequalities. This makes the formulation simpler, the solution easier to understand, and the choice of good heuristics to guide the solution strategy more straightforward.
- Secondly, although CSP algorithms are essentially very simple, they can sometimes find solution more quickly than if integer programming methods are used.

6.1.4 Constraint Solving

Constraint Solving differs from Constraint Satisfaction by using variables with infinite domains. Also, the individual constraints are more complicated, *e.g.*, nonlinear equalities. Consequently, the constraint solving algorithms uses the algebraic and numeric methods instead of combinations and search. However, there exists an approach which discretizes the infinite domain into finite number of components and, then, applies the techniques of constraint satisfaction.

6.2 Features

Constraint Programming has the features as follows:

- It is declarative and does not fix the information flow in one direction.
- The definition of constraint solution and the solution method vary according to the target area to be constrained.
- Efficiently searching is expected by using constraints.
- Prototyping is easy. Development productivity is high.
- It has a close affinity with Object-oriented Paradigm.

Declaration is the most notable feature of Constraint Programming. This feature is related to separation of the problem definition and the solution method. Functional Programming and Logic Programming also have the feature as declaration. However in Functional Programming, the information flow is fixed. The information flow in Logic Programming Language used to be fixed in point of implementation. Of course, what the flow of information can be fixed is profitable for efficiently solving problems in many cases. Besides it is not necessarily true that interactivity is available all the time. However Constraint Programming can be said to be the most thorough concerning declaration.

Moreover, the efficient problem solution is considered in point of efficiently searching by using constraint. As the definition of constraint solution and the solution method vary according to the target area to be constrained, constraint solver must be provided for every target area. Conversely, Constraint Programming can be considered very flexible and open because various efficient solutions can be gathered in the framework of Constraint Programming.

The features that prototyping is easy or development productivity is high owe a great deal to declaration. Besides when the constraint is regarded as the relation between objects, it is expected that Constraint Programming have a close affinity with object-oriented paradigm. Actually many Constraint Processing Systems are objected-oriented.

6.3 Modeling in Constraint Programming

Modeling in constraint programming revolves around the details of what is possible. For example, if you need to schedule a large number of resources and activities that respect capacity limitations, operational sequencing requirements, and business policies while meeting individual customer service goals, these modeling features are available to you:

Optional tasks

- Model activities or processes that may or may not be performed in the final schedule.
- Tasks can be grouped to match the work breakdown structure of a problem.

Precedence constraints

- Model dependencies between tasks.
- Can include a delay.
- Can be applied to group of intervals.

Expressions on interval properties

- Express typical scheduling costs such as tardiness costs, completion costs or total duration.
- The presence status of an optional interval can be used to express completion costs or resource costs.
- Can be applied to group of intervals.

Finite capacity reservoirs and resources

- Specify limits on the number of tasks that can be performed in parallel with a common resources.
- Set constraints on inventory levels.

Set-up times and batches — Calendars

- State that some tasks cannot start or end on some dates.
- State that some tasks that cannot overlap some dates.

- Define resource breaks.
- State that resource productivity changes over time.

Resource states

- These modeling features can be used for combinatorial optimization problems.
- Arithmetic linear and non-linear constraints.
- Logical constraints.

Specialized constraints and expressions.

- The all-different constraint: Enforces uniqueness for each variable in an array.
- The pack constraint: Packs items into containers with finite capacity in one dimension (time, weight, budget, and so on).
- The lexicographic constraint: Enforces a lexical ordering between groups of decision variables and is convenient to break symmetries.
- The count expression.
- The standard deviation expression.

Compatibility and incompatibility constraints

- Define possible assignments for arrays of decision variables. They can be used, for instance, to model allowed transitions in a sequencing problem.

6.3.1 Model-and-Run Constraint Programming Optimizer

Some concepts are fundamental to the way constraint programming generates workable schedules and solutions from such complexity:

- Awareness of time: CP Optimizer algorithms can determine the best sequence of activities that satisfy capacity limitations, setup times, maintenance requirements, availability restrictions, individual delivery due dates, and more by exploiting the information in the specialized scheduling and combinatorial optimization modeling constructs.
- Aggressive elimination of possibilities: Domain reduction is the internal mechanism that makes difficult allocation, sequencing, and scheduling problems manageable. Every resource or activity to be scheduled has a certain number of possible values that can be systematically reduced until a solution is reached. Every reduction makes the problem easier since the optimizer propagates the results of the decision throughout the model.
- Rapidly traversing the decision tree: The concept of systematically exploring a decision tree for feasible and efficient solutions relates to domain reduction. One benefit of this structured approach is the ability to move flexibly throughout the search space and to backtrack when early choices turn out to be dead ends.
- Adaptive Search: The default search consists of several search techniques that are dynamically changed during the search by adapting to the problem

at hand. The different search techniques include but are not restricted to LNS (large neighborhood search) and GA (genetic algorithms).

These concepts are built into the model-and-run optimizer of CP Optimizer but are also available to build a customized search.

6.3.2 The Optimizer

- Solves a large range of problems with default settings.
- Automatically removes redundant constraints.
- Reformulates the model to use constraints that are more efficiently propagated.
- Identifies conflicting constraints.
- Quickly finds feasible solutions for use in multi-model architectures such as column generation.

The optimizer is tested on an extensive library of models, and the search engine can be tuned by parameters or callbacks.

6.4 Binarization of Constraints

Most algorithms of constraint satisfaction restrict to the CSPs in which each constraint is either unary or binary. Such CSP is usually referred as a binary CSP.

Consequently, a binary CSP can be depicted by a constraint graph (sometimes referred as a constraint network), in which each node represents a variable, and each arc represents a constraint between variables represented by the end points of the arc. A unary constraint is formally represented by an arc originating and terminating at the same node. Clearly, the unary constraint can be immediately satisfied by reducing the domain of the constrained variable (node consistency), and thus, arcs representing unary constraints can be removed from the constraint network to simplify the constraint satisfaction algorithms.

Example:

constraint system:

X#Y

Y#Z

X#Z

constraint network:

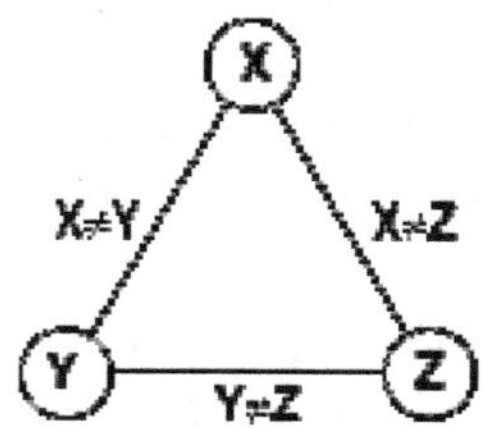

Because it is possible to convert any CSP with n-ary constraints to another equivalent binary CSP, the restriction to binary CSPs is not limitative at all.

The conversion of arbitrary CSP to an equivalent binary CSP is based on the idea of introducing a new variable that encapsulates the set of constrained variables. This newly introduced variable, we call it an **encapsulated variable**, has assigned a domain that is a Cartesian product of the domains of individual variables. Note, that if the domains of individual variables are finite than the Cartesian product of the domains, and thus the resulting domain, is still finite.

Example:

original (individual) variables and their domains:

X::[1,2]

Y::[3,4]

Z::[5,6]

encapsulated variable and its domain:

U:: [(1,3,5),(1,3,6),
(1,4,5),(1,4,6),
(2,3,5),(2,3,6)
(2,4,5),(2,4,6)]

Now, arbitrary n-ary constraint can be converted to equivalent unary constraint that constrains the variable which appears as an encapsulation of the original individual variables. As we mentioned above, this unary constraint can be immediately satisfied by reducing the domain of encapsulated variable. Briefly speaking, n-ary constraint can be substituted by an encapsulated variable with the domain corresponding to the constraint.

Example:

original constraint and variables:

X+Y=Z

X::[1,2]; Y::[3,4]; Z::[5,6]

encapsulated variable and reduced domain:

U::[(1,4,5),(2,3,5),(2,4,6)]

In fact, this transformation solves individual constraints. Now, it remains to combine these individual solutions to the solution of the constraint system. There are two ways how to bind the encapsulated variables:

6.4.1 With original variables

Combination of the original and encapsulated variables naturally expresses the original non-binary CSP as the encapsulated variables directly correspond to the original constraints. Newly introduced constraints just bind the original and encapsulated variables in a following way:

X=i_th_argument_of(U),

where X is the original variable, U is the encapsulated variable and i is the "position of X within U".

Example:

original (non-binary) CSP:

X+Y=Z, X<Y

X::[1,2]; Y::[3,4]; Z::[5,6]

equivalent binary CSP:

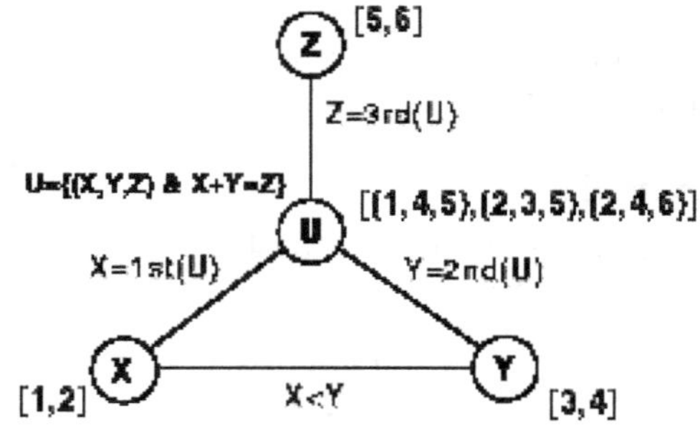

The advantage of this approach is the preservation of the original variables, so the solution of the original CSP can be directly obtained from the solution of the binary CSP (the evaluation of the encapsulated variables is omitted). Another advantage is the preservation of the binary constraints from the original CSP.

6.4.2 Without original Variables

The other approach to represent converted binary CSP is based on using encapsulated variables only. Then, the newly introduced constraints just binds these variables via common components in a following way:

i_th_argument_of(U)=j_th_argument_of(V),

where U and V are encapsulated variables and i and j respectively are the "positions of common component".

Example:

original (non-binary) CSP:

X+Y=Z, X<Y

X::[1,2]; Y::[3,4]; Z::[5,6]

equivalent binary CSP:

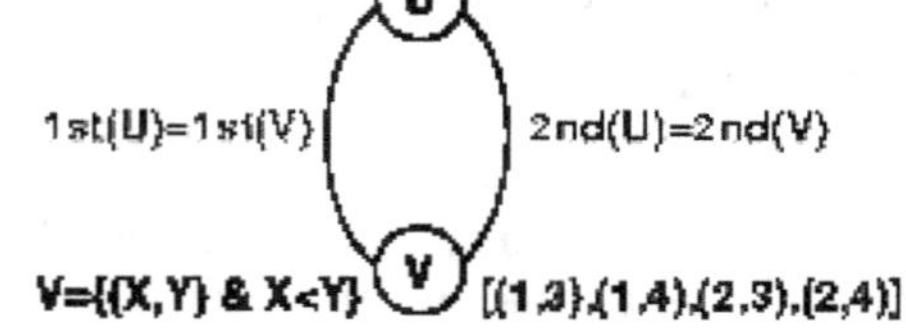

In this approach, each constraint from the original CSP is represented by an encapsulated variable. The resulting constraint network is smaller than the network constructed by the previous method, however, the solution, *i.e.*, the valuation of original variables, has to be extracted from the valuation of encapsulated variables.

6.5 Systematic Search Algorithms

Most algorithms for solving CSPs search systematically through the possible assignments of values to variables. Such algorithms are guaranteed to find a solution, if one exists, or to prove that the problem is insoluble. The disadvantage of these algorithms is that they take a very long time to do so.

6.5.1 Generate and Test (GT)

Generate-and-test method originates from the mathematical approach to solving combinatorial problems. First, the GT algorithm guesses the solution and, then, it tests whether this solution is correct, *i.e.*, whether the solution satisfies the original constraints. In this paradigm, each possible combination of the variable assignments is systematically generated and tested to see if it satisfies all the constraints. The first combination that satisfies all the constraints is the solution. The number of combinations considered by this method is the size of the Cartesian product of all the variable domains.

Algorithm GT

```
gt(Variables,Constraints,Solution):-
generate(Variables,Solution),
test(Constraints,Solution).
generate([V::D|RemainingVariables],[V-X|PartialSolution]):-
select_value(X,D),
generate(RemainingVariables,PartialSolution).
generate([],[]).
test([C|RemainingConstraints],Solution):-
test_constraint(C,Solution),
test(RemainingConstraints,Solution).
test([],_).
```

V::D denotes variable V and its domain D

The generate-and-test approach is not very efficient because it generates many wrong assignments of values to variables which are rejected in the testing phase. In addition, the generator leaves out the conflicting instantiations and it generates other assignments independently of the conflict. Visibly, one can get far better efficiency, if the validity of the constraint is tested as soon as its respective variables are instantiated. In fact, this method is used by the backtracking approach.

6.5.1.1 Backtracking (BT)

The most common algorithm for performing systematic search is backtracking. Backtracking incrementally attempts to extend a partial solution that specifies consistent values for some of the variables, toward a complete solution, by repeatedly choosing a value for another variable consistent with the values in the current partial solution.

Backtracking can be seen as a merge of generate and test phases from the GT approach. In the BT method, variables are instantiated sequentially and as soon as all the variables relevant to a constraint are instantiated, the validity of the constraint is checked. If a partial solution violates any of the constraints, backtracking is performed to the most recently instantiated variable that still has alternatives available. Clearly, whenever a partial instantiation violates a constraint, backtracking is able to eliminate a subspace from the Cartesian product of all variable domains. Consequently, backtracking is strictly better than generate-and-test, however, its running complexity for most nontrivial problems is still exponential.

Algorithm BT

```
bt([V::D | RemainingVariables],Constraints,PartialSolution,Solution):-
select_value(X,D),
NewPartialSolution=[V-X | PartialSolution],
test(Constraints,NewPartialSolution,RemainingConstraints),
bt(RemainingVariables,RemainingConstraints,NewPartialSolution,Solution).
bt([],[],Solution,Solution).
test([C | RemainingConstraints],PartialSolution,NonTestedConstraints):-
(can_be_tested(C,PartialSolution)
-> test_constraint(C,PartialSolution),NonTestedConstraints=Constraints
; NonTestedConstraints=[C | Constraints]),
test(RemainingConstraints,PartialSolution,Constraints).
test([],_,[]).
can_be_tested(Constraint,PartialSolution):-
variables(Constraint,Variables),
are_instantiated(Variables,PartialSolution).
```

There are three major drawbacks of the standard backtracking scheme. One is thrashing, *i.e.*, repeated failure due to the same reason. Thrashing occurs because the standard backtracking algorithm does not identify the real reason of the conflict, *i.e.*, the conflicting variables. Therefore, search in different parts of the space keeps failing for the same reason. Thrashing can be avoided by intelligent backtracking, *i.e.*, by a scheme on which backtracking is done directly to the variable that caused the failure.

The other drawback of backtracking is having to perform redundant work. Even if the conflicting values of variables is identified during the intelligent backtracking, they are not remembered for immediate detection of the same conflict in a subsequent computation. There is a backtracking based method that eliminates both of the above drawbacks of backtracking. This method is traditionally called dependency-directed backtracking and is used in truth maintenance systems.

It should be noted that using advanced techniques adds other expenses to the algorithm that has to be balanced with the overall advantage of using them.

Finally, the basic backtracking algorithm still detects the conflict too late as it is not able to detect the conflict before the conflict really occurs, *i.e.*, after assigning the values to the all variables of the conflicting constraint. This drawback can be avoided by applying consistency techniques to forward check the possible conflicts.

6.6 Consistency Techniques

Consistency techniques were first introduced for improving the efficiency of picture recognition programs, by researchers in artificial intelligence. Picture recognition involves labelling all the lines in a picture in a consistent way. The number of possible combinations can be huge, while only very few are consistent. Consistency techniques effectively rule out many inconsistent labellings at a very early stage, and thus cut short the search for consistent labellings. These techniques have since proved to be effective on a wide variety of hard search problems.

Notice that consistency techniques are deterministic, as opposed to the search which is non-deterministic. Thus the deterministic computation is performed as soon as possible and non-deterministic computation during search is used only when there is no more propagation to done. Nevertheless, the consistency techniques are rarely used alone to solve constraint satisfaction problem completely.

In binary CSPs, various consistency techniques for constraint graphs were introduced to prune the search space. The consistency-enforcing algorithm makes any partial solution of a small sub-network extensible to some surrounding network. Thus, the potential inconsistency is detected as soon as possible.

6.6.1 Node Consistency

The simplest consistency technique is referred to as node consistency and we mentioned it in the section on binarization of constraints. The node representing a variable V in constraint graph is **node consistent** if for every value x in the current domain of V, each unary constraint on V is satisfied.

If the domain D of a variable V contains a value "a" that does not satisfy the unary constraint on V, then the instantiation of V to "a" will always result in immediate failure. Thus, the node inconsistency can be eliminated by simply removing those values from the domain D of each variable V that do not satisfy unary constraint on V.

Algorithm NC

procedure NC

for each V in nodes(G)

for each X in the domain D of V

```
if any unary constraint on V is inconsistent with X
then
delete X from D;
endif
endfor
endfor
end NC
```

6.6.2 Arc Consistency

If the constraint graph is node consistent then unary constraints can be removed because they all are satisfied. As we are working with the binary CSP, there remains to ensure consistency of binary constraints. In the constraint graph, binary constraint corresponds to arc, therefore this type of consistency is called arc consistency.

Arc (V_i,V_j) is **arc consistent** if for every value x the current domain of V_i there is some value y in the domain of V_j such that V_i=x and V_j=y is permitted by the binary constraint between V_i and V_j. Note, that the concept of arc-consistency is directional, *i.e.*, if an arc (V_i,V_j) is consistent, than it does not automatically mean that (V_j,V_i) is also consistent.

Clearly, an arc (V_i,V_j) can be made consistent by simply deleting those values from the domain of V_i for which there does not exist corresponding value in the domain of D_j such that the binary constraint between V_i and V_j is satisfied (note, that deleting of such values does not eliminate any solution of the original CSP). The following algorithm does precisely that:

Algorithm REVISE

```
procedure REVISE(Vi,Vj)
DELETE <- false;
for each X in Di do
if there is no such Y in Dj such that (X,Y) is consistent,
then
delete X from Di;
DELETE <- true;
endif;
endfor;
return DELETE;
end REVISE
```

To make every arc of the constraint graph consistent, it is not sufficient to execute REVISE for each arc just once. Once REVISE reduces the domain of some variable V_i, then each previously revised arc (V_j,V_i) has to be revised again, because some of the members of the domain of V_j may no longer be compatible with any remaining members of the revised domain of V_i. The following algorithm, known as **AC-1**, does precisely that:

Algorithm AC-1

```
procedure AC-1
Q <- {(Vi,Vj) in arcs(G),i#j};
repeat
CHANGE <- false;
for each (Vi,Vj) in Q do
CHANGE <- REVISE(Vi,Vj) or CHANGE;
endfor
until not(CHANGE)
end AC-1
```

This algorithm is not very efficient because the successful revision of even one arc in some iteration forces all the arcs to be revised again in the next iteration, even though only a small number of them are really affected by this revision. Visibly, the only arcs affected by the reduction of the domain of V_k are the arcs (V_i,V_k). Also, if we revise the arc (V_k,V_m) and the domain of V_k is reduced, it is not necessary to re-revise the arc (V_m,V_k) because non of the elements deleted from the domain of V_k provided support for any value in the current domain of V_m. The following variation of arc consistency algorithm, called AC-3, removes this drawback of AC-1 and performs re-revision only for those arcs that are possibly affected by a previous revision.

Algorithm AC-3

```
procedure AC-3
Q <- {(Vi,Vj) in arcs(G),i#j};
while not Q empty
select and delete any arc (Vk,Vm) from Q;
if REVISE(Vk,Vm) then
Q <- Q union {(Vi,Vk) such that (Vi,Vk) in arcs(G),i#k,i#m}
endif
endwhile
end AC-3
```

When the algorithm AC-3 revises the edge for the second time it re-tests many pairs of values which are already known (from the previous iteration) to be consistent or inconsistent respectively and which are not affected by the reduction of the domain. As this is a source of potential inefficiency, the algorithm **AC-4** was introduced to refine handling of edges (constraints). The algorithm works with individual pairs of values as the following example shows.

Example

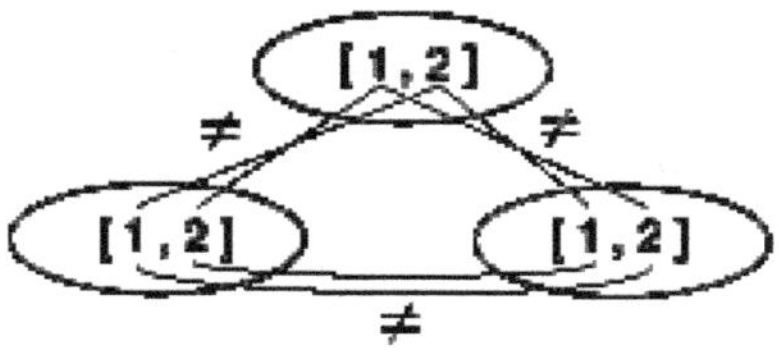

First, the algorithm AC-4 initializes its internal structures which are used to remember pairs of consistent (inconsistent) values of incidental variables (nodes) - structure $S_{i,a}$. This initialization also counts "supporting" values from the domain of incindental variable - structure $counter_{(i,j),a}$ - and it removes those values which have no support. Once the value is removed from the domain, the algorithm adds the pair <Variable,Value> to the list Q for re-revision of affected values of corresponding variables.

Algorithm INITIALIZE

```
procedure INITIALIZE
Q <- {};
S <- {}; % initialize each element of structure S
for each (Vi,Vj) in arcs(G) do % (Vi,Vj) and (Vj,Vi) are same elements
for each a in Di do
total <- 0;
for each b in Dj do
if (a,b) is consistent according to the constraint (Vi,Vj) then
total <- total+1;
Sj,b <- Sj,b union {<i,a>};
endif
endfor;
counter[(i,j),a] <- total;
if counter[(i,j),a]=0 then
delete a from Di;
```

```
Q <- Q union {<i,a>};
endif;
endfor;
endfor;
return Q;
end INITIALIZE
```

After the initialization, the algorithm AC-4 performs re-revision only for those pairs of values of incindental variables that are affected by a previous revision.

Algorithm AC-4

```
procedure AC-4
Q <- INITIALIZE;
while not Q empty
select and delete any pair <j,b> from Q;
for each <i,a> from Sj,b do
counter[(i,j),a] <- counter[(i,j),a] - 1;
if counter[(i,j),a]=0 & a is still in Di then
delete a from Di;
Q <- Q union {<i,a>};
endif
endfor
endwhile
end AC-4
```

Both algorithms, AC-3 and AC-4, belong to the most widely used algorithms for maintaining arc consistency. It should be also noted that there exist other algorithms AC-5, AC-6, AC-7, *etc.* but their are not used as frequently as AC-3 or AC-4.

Maintaining arc consistency removes many inconsistencies from the constraint graph but is any (complete) instantiation of variables from current (reduced) domains a solution to the CSP? If the domain size of each variable becomes one, then the CSP has exactly one solution which is obtained by assigning to each variable the only possible value in its domain. Otherwise, the answer is no in general. The following example shows such a case where the constraint graph is arc consistent, domains are not empty but there is still no solution satisfying all constraints.

Example:

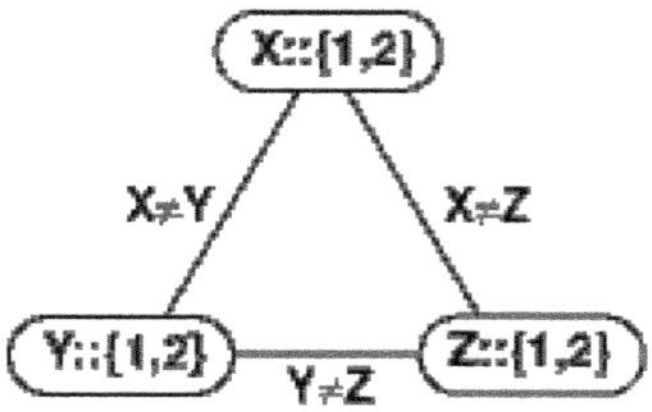

This constraint graph is arc consistent but there is no solution that satisfies all the constraints.

6.6.3 K-Consistency (Path Consistency)

Given that arc consistency is not enough to eliminate the need for backtracking, is there another stronger degree of consistency that may eliminate the need for search? The above example shows that if one extends the consistency test to two or more arcs, more inconsistent values can be removed.

A graph is **K-consistent** if the following is true: Choose values of any K-1 variables that satisfy all the constraints among these variables and choose any Kth variable. Then there exists a value for this Kth variable that satisfies all the constraints among these K variables. A graph is **strongly K-consistent** if it is J-consistent for all J<=K.

Node consistency is equivalent to strong 1-consistency and arc-consistency is equivalent to strong 2-consistency (arc-consistency is usually assumed to include node-consistency as well). Algorithms exist for making a constraint graph strongly K-consistent for K>2 but in practice they are rarely used because of efficiency issues. The exception is the algorithm for making a constraint graph strongly 3-consistent that is usually referred as **path consistency**. Nevertheless, even this algorithm is too hungry and a weak form of path consistency was introduced.

A node representing variable V_i is **restricted path consistent** if it is arc-consistent, *i.e.*, all arcs from this node are arc-consistent, and the following is true: For every value a in the domain D_i of the variable V_i that has *just one supporting value* b from the domain of incidental variable V_j there exists a value c in the domain of other incidental variable V_k such that (a,c) is permitted by the binary constraint between V_i and V_k, and (c,b) is permitted by the binary constraint between V_k and V_j.

The algorithm for making graph restricted path consistent can be naturally based on AC-4 algorithm that counts the number of supporting values. Although this algorithm removes more inconsistent values than any arc-consistency algorithm it does not eliminate the need for search in general. Clearly, if a constraint graph containing n nodes is strongly n-consistent, then a solution to the CSP can be found without any search. But the worst-case complexity of the algorithm for obtaining n-consistency in a n-node constraint graph is also exponential. If the graph is (strongly) K-consistent for K<n, then in general, backtracking cannot be avoided, *i.e.*, there still exist inconsistent values.

6.7 Constraint Propagation

As seen, there are two rather different schemes for solving CSP: backtracking and consistency techniques where as the third possible scheme embeds a consistency algorithm inside a backtracking algorithm as follows.

As a skeleton we use the simple backtracking algorithm that incrementally instantiates variables and extends a partial solution that specifies consistent values for some of the variables, toward a complete solution, by repeatedly choosing a value for another variable. After assigning a value to the variable, some consistency technique is applied to the constraint graph. Depending on the degree of consistency technique we get various constraint satisfaction algorithms.

6.7.1 Backtracking

Even simple backtracking (BT) performs some kind of consistency technique and it can be seen as a combination of pure generate & test and a fraction of arc consistency. The BT algorithm tests arc consistency among already instantiated variables, *i.e.*, the algorithm checks the validity of constraints considering the partial instantiation. Because the domains of instantiated variables contains just one value, it is possible to check only those constraints/arcs containing the last instantiated variable. If any domain is reduced then the corresponding constraint is not consistent and the algorithm backtracks to a new instantiation. The following procedure AC3-BT is called each time a new value is assigned to some variable Vcv (cv is the consecutive number of the variable in the order of instantiating variables).

Algorithm AC-3 for Backtracking

```
procedure AC3-BT(cv)
Q <- {(Vi,Vcv) in arcs(G),i<cv};
consistent <- true;
while not Q empty & consistent
select and delete any arc (Vk,Vm) from Q;
consistent <- REVISE(Vk,Vm)
endwhile
return consistent
end AC3-BT
```

The BT algorithm detects the inconsistency as soon as it appears and, therefore, it is far away efficient than the simple generate & test approach. But it has still to perform too much search.

Example:

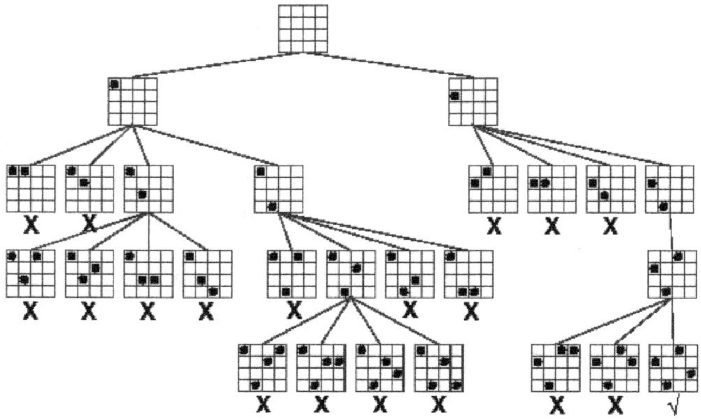

Figure 6.1: Queens Problem and BT.

The BT algorithm can be easily extended to backtrack to the conflicting variable and, thus, to incorporate some form of look-back scheme or intelligent backtracking. Nevertheless, this adds some additional expenses to the algorithm and it seems that preventing possible future conflicts is more reasonable than recovering from them.

6.7.2 Forward Checking

Forward checking is the easiest way to prevent future conflicts. Instead of performing arc consistency to the instantiated variables, it performs restricted form of arc consistency to the not yet instantiated variables. We speak about restricted arc consistency because forward checking checks only the constraints between the current variable and the future variables. When a value is assigned to the current variable, any value in the domain of a "future" variable which conflicts with this assignment is (temporarily) removed from the domain. The advantage of this is that if the domain of a future variable becomes empty, it is known immediately that the current partial solution is inconsistent. Forward checking therefore allows branches of the search tree that will lead to failure to be pruned earlier than with simple backtracking. Note that whenever a new variable is considered, all its remaining values are guaranteed to be consistent with the past variables, so the checking an assignment against the past assignments is no longer necessary.

Algorithm AC-3 for Forward Checking

```
procedure AC3-FC(cv)
Q <- {(Vi,Vcv) in arcs(G),i>cv};
consistent <- true;
while not Q empty & consistent
select and delete any arc (Vk,Vm) from Q;
if REVISE(Vk,Vm) then
```

consistent <- not Dk empty

endif

endwhile

return consistent

end AC3-FC

Forward checking detects the inconsistency earlier than simple backtracking and thus it allows branches of the search tree that will lead to failure to be pruned earlier than with simple backtracking. This reduces the search tree and the overall amount of work done. But it should be noted that forward checking does more work when each assignment is added to the current partial solution.

Example:

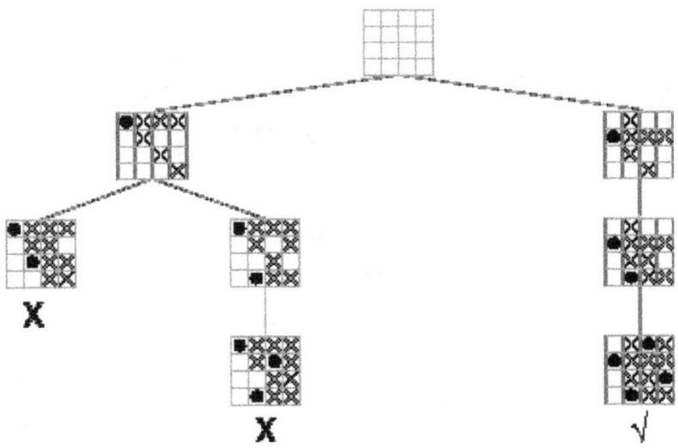

Figure 6.2: Queens Problem and FC.

Forward checking is almost always a much better choice than simple backtracking.

6.7.3 Look Ahead

Forward checking checks only the constraints between the current variable and the future variables. So why not to perform full arc consistency that will further reduces the domains and removes possible conflicts? This approach is called (full) look ahead or maintaining arc consistency (MAC).

The advantage of look ahead is that it detects also the conflicts between future variables and therefore allows branches of the search tree that will lead to failure to be pruned earlier than with forward checking. Also as with forward checking, whenever a new variable is considered, all its remaining values are guaranteed to be consistent with the past variables, so the checking an assignment against the past assignments is no necessary.

Algorithm AC-3 for Look Ahead

procedure AC3-LA(cv)

Q <- {(Vi,Vcv) in arcs(G),i>cv};

```
consistent <- true;
while not Q empty & consistent
select and delete any arc (Vk,Vm) from Q;
if REVISE(Vk,Vm) then
Q <- Q union {(Vi,Vk) such that (Vi,Vk) in arcs(G),i#k,i#m,i>cv}
consistent <- not Dk empty
endif
endwhile
return consistent
end AC3-LA
```

Look ahead prunes the search tree further more than forward checking but, again, it should be noted that look ahead does even more work when each assignment is added to the current partial solution than forward checking.

Example:

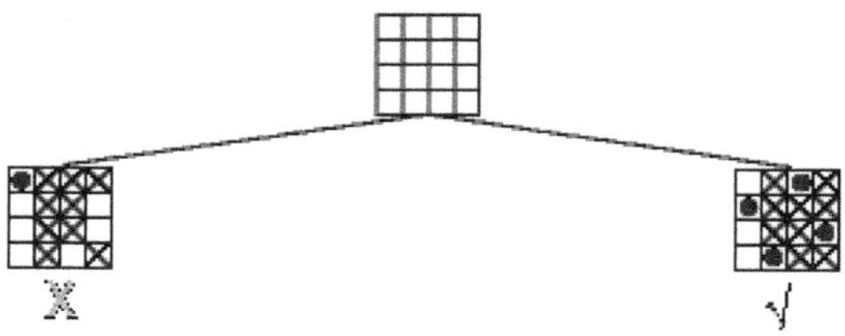

Figure 6.3: Queens Problem and LA.

The following figure shows which constraints are tested when the above described propagation techniques are applied.

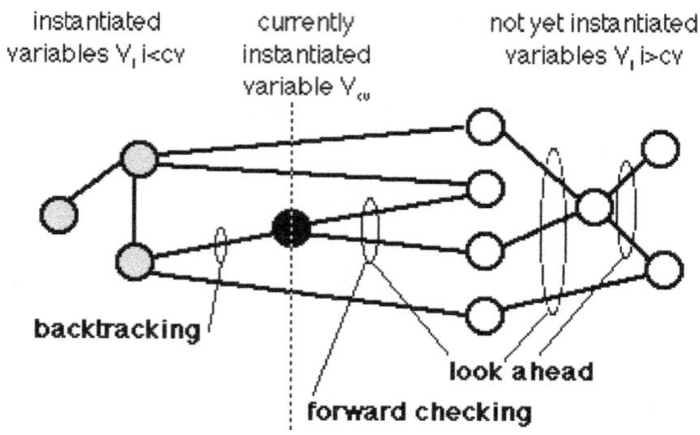

Figure 6.4: Comparison of Propagation Techniques.

6.8 Value and Variable Ordering

A search algorithm for constraint satisfaction requires the order in which variables are to be considered to be specified as well as the order in which the values are assigned to the variable on backtracking. Choosing the right order of variables (and values) can noticeably improve the efficiency of constraint satisfaction.

6.8.1 Variable Ordering

Experiments and analysis of several researchers have shown that the ordering in which variables are chosen for instantiation can have substantial impact on the complexity of backtrack search. The ordering may be either a static ordering, in which the order of the variables is specified before the search begins, and it is not changed thereafter, or a dynamic ordering, in which the choice of next variable to be considered at any point depends on the current state of the search.

Dynamic ordering is not feasible for all search algorithms, *e.g.*, with simple backtracking there is no extra information available during the search that could be used to make a different choice of ordering from the initial ordering. However, with forward checking, the current state includes the domains of the variables as they have been pruned by the current set of instantiations, and so it is possible to base the choice of next variable on this information.

Several heuristics have been developed and analyzed for selecting variable ordering. The most common one is based on the "first-fail" principle, which can be explained as "To succeed, try first where you are most likely to fail."

In this method, the variable with the fewest possible remaining alternatives is selected for instantiation. Thus the order of variable instantiations is, in general, different in different branches of the tree, and is determined dynamically. This method is based on assumption that any value is equally likely to participate in a solution, so that the more values there are, the more likely it is that one of them will be a successful one.

The first-fail principle may seem slightly misleading; after all, we do not want to fail. The reason is that if the current partial solution does not lead to a complete solution, then the sooner we discover this better. Hence encouraging early failure, if failure is inevitable, is beneficial in the long term. On the other end, if the current partial solution can be extended to a complete solution, then every remaining variable must be instantiated and the one with smallest domain is likely to be the most difficult to find a value for (instantiating other variables first may further reduce its domain and lead to a failure). Hence the principle could equally well be stated as:

"Deal with hard cases first: they can only get more difficult if you put them off."

This heuristic should reduce the average depth of branches in the search tree by triggering early failure. Another heuristic, that is applied when all variables have the same number of values, is to choose the variable which participates in most constraints (in the absence of more specific information on which constraints are likely to be difficult to satisfy, for instance). This heuristic follows also the principle of dealing with hard cases first.

There is also a heuristic for static ordering of variables that is suitable for simple backtracking. This heuristic says: choose the variable which has the largest number of constraints with the past variables. For instance, during solving graph coloring problem, it is reasonable to assign color to the vertex which has common arcs with already colored vertices so the conflict is detected as soon as possible.

6.8.2 Value Ordering

Once the decision is made to instantiate a variable, it may have several values available. Again, the order in which these values are considered can have substantial impact on the time to find the first solution. However, if all solutions are required or there are no solutions, then the value ordering is indifferent.

A different value ordering will rearrange the branches emanating from each node of the search tree. This is an advantage if it ensures that a branch which leads to a solution is searched earlier than branches which lead to death ends. For example, if the CSP has a solution, and if a correct value is chosen for each variable, then a solution can be found without any backtracking.

Suppose we have selected a variable to instantiate: how should we choose which value to try first? It may be that none of the values will succeed, in that case, every value for the current variable will eventually have to be considered, and the order does not matter. On the other hand, if we can find a complete solution based on the past instantiations, we want to choose a value which is likely to succeed, and unlikely to lead to a conflict. So, we apply the "succeed first" principle.

One possible heuristic is to prefer those values that maximize the number of options available. Visibly, the algorithm AC-4 is good for using this heuristic as it counts the supporting values. It is possible to count "promise" of each value, that is the product of the domain sizes of the future variables after choosing this value (this is an upper bound on the number of possible solutions resulting from the assignment). The value with highest promise should be chosen. It is also possible to calculate the percentage of values in future domains which will no longer be usable. The best choice would be the value with lowest cost.

Another heuristic is to prefer the value (from those available) that leads to an easiest to solve CSP. This requires to estimate the difficulty of solving a CSP. One method propose to convert a CSP into a tree-structured CSP by deleting a minimum number of arcs and then to find all solutions of the resulting CSP (higher the solution count, easier the CSP).

For randomly generated problems, and probably in general, the work involved in assessing each value is not worth the benefit of choosing a value which will on average be more likely to lead to a solution than the default choice. In particular problems, on the other hand, there may be information available which allows the values to be ordered according to the principle of choosing first those most likely to succeed.

6.9 Reducing Search

Ii is seen that Value and Variable Ordering carries orderings which can noticeably improve the efficiency of backtrack search. The question is that does there exist any variable ordering which can eliminate the need for backtracking, for this, lets know certain terms.

6.9.1 Background

An ordered constraint graph is a constraint graph whose vertices have been ordered linearly. The width of the vertex in an ordered constraint graph is the number of constraint arcs that lead from the vertex to the previous vertices (in the linear order). The width of the ordered constraint graph is the maximum width of any of its vertices and the width of the constraint graph is the minimum width of all the orderings of that graph. In general, the width of a constraint graph depends upon its structure.

Example:

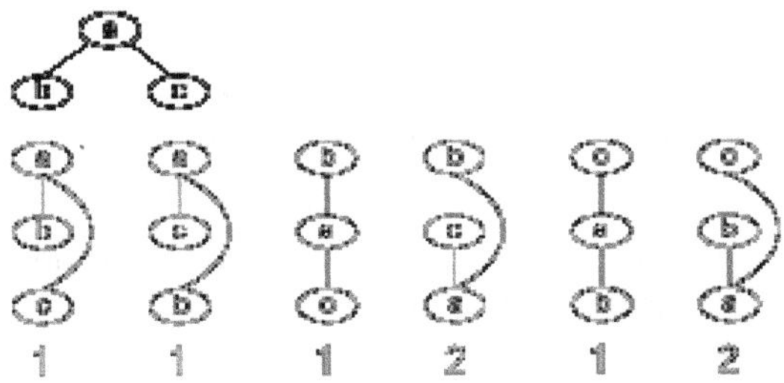

constraint graph

different ordered constraint graphs

widths of the ordered constraint graphs

The width of the constraint graph is 1.

6.9.2 Theorem

If a constraint graph is strongly K-consistent, and K>w where w is the width of the constraint graph, then there exists a search order that is backtrack free.

The proof of the above theorem is straightforward. There exists an ordering of the graph such that the number of constraint arcs leading from any vertex of the graph to the previous vertices is at most w. Now if the variables are instantiated using this ordering, then whenever a new variable is instantiated, a value for this variable is consistent with all the previous assignments because: (i) this value is to be consistent with the assignments of at most w other variables, and (ii) the graph is strongly (w+1)-consistent.

Interestingly, all tree structured constrained graphs have width 1, so it is possible to find at least one ordering that can be used to solve the constraint graph without backtracking provided that the constraint graph is arc consistent.

6.9.3 Cycle-Cutset

According to the above theorem, acyclic constraint graphs are globally consistent iff they are arc consistent. So, if we first instantiate consistently the nodes/variables which "cut" all the cycles in a graph (the set of such nodes is called cycle cutset) and make the graph arc consistent then it is possible to find ordering of rest variables that eliminate the need for backtracking. This observation makes the basis of the cycle cutset algorithm that solves the cycle cutset independently and then extends this partial solution to a complete solution in a backtrack-free manner using the variable ordering described in the above section.

Algorithm Cycle-Cutset

1. partition the variables into two sets:

   a cycle-cutset C, and T, the complement of C;
2. find a(nother) solution to the the problem with variables C only

   if no solution can be found then return fail
3. enforce directed arc consistency from C to T

   if any domain is empty then backtrack to 2
4. use backtrack-free search for extending the partial solution to a complete solution.

The major problem with the cycle cutset method is its potential for thrashing, *i.e.*, occurrence of backtracks in the step 3 of above algorithm.

6.9.4 MACE (MAC Extended)

MACE is an extended version of the popular algorithm MAC that combines two ideas: instantiate less and propagate less. MACE instantiates only a subset of the CSP variables while maintaining only a partial arc consistent state of the constraint network. The gain in efficiency is twofold. Instantiating a smaller number of variables aims at reducing the number of backtracks (and, accordingly, the number of constraint-checks, nodes visited, values deleted, *etc.*). MACE uses the idea of cycle cutset of a constraint graph while removes its major drawback, thrashing, by maintaining arc consistency.

Algorithm MACE

1. enforce arc consistency (if it is not possible then return fail)
2. partition the variables into two sets:

   a cycle-cutset C and the set of variables which are not involved in any cycle U disconnect U from the graph
3. while C#0 do

   3a) set value to a variable in C

   3b) enforce arc consistency

   if failed then backtrack to 3a

   3c) disconnect singleton variables (add them to U)

   3d) disconnect "cycle-free" variables (add them to U)

4. reconnect variables in U

   enforce directed arc consistency from C to U

5. conduct backtrack-free search for a complete solution

6.9.5 Heuristics

MACE and CC need an algorithm to find a cycle-cut. There is no known polynomial algorithm for finding the minimum cycle-cutset. Fortunately, there are several heuristics which yield a good cycle-cutset at a reasonable cost.

- The simplest heuristic sorts first the variable in decreasing order of their degree. Then, starting with the variable with the highest degree, as long as the graph still has cycles, add the variable to the cycle-cutset and remove it, together with all the edges involving it, from the graph.
- A smaller cutset can be obtained if, before adding a variable to the cutset, we check whether it is part of any cycle or not.
- The third heuristic determines dynamically the number of cycles in which each variable is involved and adds to the cutset at each step the variable participating in the most cycles.

Visibly, second and third heuristic yields smaller cutsets than the first heuristic. Based on real-life observations, the third heuristic yields slightly smaller cutsets which translate sometimes into small gains in efficiency, but no major improvement upon the second heuristic. In case of large problems probably the second heuristic is the best choice, because of its lower cost.

6.10 Heuristics and Stochastic Algorithms

As observed, greedy local search strategies became popular, again where such algorithms incrementally alter inconsistent value assignments to all the variables by using a "repair" or "hill climbing" metaphor to move towards more and more complete solutions. To avoid getting stuck at "local optima" they are equipped with various heuristics for randomizing the search. Their stochastic nature generally voids the guarantee of "completeness" provided by the systematic search methods.

The local search methodology uses the following terms:

- state (configuration): one possible assignment of all variables; the number of states is equal to the product of domains' sizes.
- evaluation value: the number of constraint violations of the state (sometimes weighted).
- neighbor: the state which is obtained from the current state by changing one variable value.
- local-minimum: the state that is not a solution and the evaluation values of all of its neighbors are larger than or equal to the evaluation value of this state.

- strict local-minimum: the state that is not a solution and the evaluation values of all of its neighbors are larger than the evaluation value of this state.
- non-strict local-minimum: the state that is a local-minimum but not a strict local-minimum.

6.10.1 Hill-Climbing

Hill-climbing is probably the most known algorithm of local search. The idea of hill-climbing is:

- start at randomly generated state.
- move to the neighbor with the best evaluation value.
- if a strict local-minimum is reached then restart at other randomly generated state.

This procedure repeats till the solution is found. In the algorithm, that we present here, the parameter Max_Flips is used to limit the maximal number of moves between restarts which helps to leave non-strict local-minimum.

Algorithm Hill-Climbing

```
procedure hill-climbing(Max_Flips)
restart: s <- random valuation of variables;
for j:=1 to Max_Flips do
if eval(s)=0 then return s endif;
if s is a strict local minimum then
goto restart
else
s <- neighborhood with smallest evaluation value
endif
endfor
goto restart
end hill-climbing
```

Note, that the hill-climbing algorithm has to explore all neighbors of the current state before choosing the move. This can take a lot of time.

6.10.2 Min-Conflicts

To avoid exploring all neighbors of the current state some heuristics were proposed to find a next move. Min-conflicts heuristics chooses randomly any conflicting variable, *i.e.*, the variable that is involved in any unsatisfied constraint, and then picks a value which minimizes the number of violated constraints (break ties randomly). If no such value exists, it picks randomly one value that does not increase the number of violated constraints (the current value of the variable is picked only if all the other values increase the number of violated constraints).

Algorithm Min-Conflicts

```
procedure MC(Max_Moves)
s <- random valuation of variables;
nb_moves <- 0;
while eval(s)>0 & nb_moves<Max_Moves do
choose randomly a variable V in conflict;
choose a value v' that minimizes the number of conflicts for V;
if v' # current value of V then
assign v' to V;
nb_moves <- nb_moves+1;
endif
endwhile
return s
end MC
```

Note, that the pure min-conflicts algorithm presented above is not able to leave local-minimum. In addition, if the algorithm achieves a strict local-minimum it does not perform any move at all and, consequently, it does not terminate.

6.10.3 Min-Conflicts-Random-Walk

Because the pure min-conflicts algorithm cannot go beyond a local-minimum, some noise strategies were introduced in MC. Among them, the random-walk strategy becomes one of the most popular. For a given conflicting variable, the random-walk strategy picks randomly a value with probability p, and apply the MC heuristic with probability 1-p.

Algorithm Min-Conflicts-Random-Walk

```
procedure MCRW(Max_Moves,p)
s <- random valuation of variables;
nb_moves <- 0;
while eval(s)>0 & nb_moves<Max_Moves do
if probability p verified then
choose randomly a variable V in conflict;
```

```
choose randomly a value v' for V;
else
choose randomly a variable V in conflict;
choose a value v' that minimizes the number of conflicts for V;
endif
if v' # current value of V then
assign v' to V;
nb_moves <- nb_moves+1;
endif
endwhile
return s
end MCRW
```

This algorithm is controlled by the random probability p, it should be clear that the value for this parameter has a big influence on the performance of the algorithm. The preliminary studies determined the following feasible ranges of parameter values 0.02 <= p <= 0.1.

6.10.4 Steepest-Descent-Random-Walk

Stepest-Descent algorithm is a hill-climbing version of the min-conflicts algorithm. Instead of selecting the variable in conflict randomly, this algorithm explores the whole neighborhood of the current configuration and selects the best neighbor according to the evaluation value. Again, the algorithm can be randomized by using random-walk strategy to avoid getting stuck at "local optima"

Algorithm Steepest-Descent-Random-Walk

```
procedure SDRW(Max_Moves,p)
s <- random valuation of variables;
nb_moves <- 0;
while eval(s)>0 & nb_moves<Max_Moves do
if probability p verified then
choose randomly a variable V in conflict;
choose randomly a value v' for V;
else
choose a move <V,v'> with the best performance
endif
```

```
if v' # current value of V then
assign v' to V;
nb_moves <- nb_moves+1;
endif
endwhile
return s
end SDRW
```

6.10.5 Tabu-Search

Tabu search (TS) is another method to avoid cycling and getting trapped in local minimum. It is based on the notion of tabu list, that is a special short term memory that maintains a selective history, composed of previously encountered configurations or more generally pertinent attributes of such configurations (we use a couple <Variable, Value> as the characterization of the configuration). A simple TS strategy consist in preventing configurations of tabu list from being recognized for the next k iterations (k, called tabu tenue, is the size of tabu list). Such a strategy prevents Tabu from being trapped in short-term cycling and allows the search process to go beyond local optima.

Tabu restrictions may be overridden under certain conditions, called aspiration criteria. Aspiration criteria define rules that govern whether next configuration is considered as a possible move even it is tabu. One widely used aspiration criterion consists of removing a tabu classification from a move when the move leads to a solution better than that obtained so far.

Algorithm Tabu-Search

```
procedure tabu-search(Max_Iter)
s <- random valuation of variables;
nb_iter <- 0;
initialize randomly the tabu list;
while eval(s)>0 & nb_iter<Max_Iter do
choose a move <V,v'> with the best performance among the non-tabu moves
and the moves satisfying the aspiration criteria;
introduce <V,v> in the tabu list, where v is the current values of V;
remove the oldest move from the tabu list;
assign v' to V;
nb_iter <- nb_iter+1;
endwhile
```

return s

end tabu-search

Again, the performance of Tabu Search is greatly influenced by the size of tabu list tl. A preliminary studies determined the following feasible range of parameter values 10 <= tl <= 35.

6.10.6 GSAT

GSAT is a greedy local search procedure for satisfying logic formulas in a conjunctive normal form (CNF). Such problems are called SAT or k-SAT (k is a number of literals in each clause of the formula) and are known to be NP-c (each NP-hard problem can be transformed to NP-complex problem).

The procedure starts with an arbitrary instantiation of the problem variables and offers to reach the highest satisfaction degree by succession of small transformations called repairs or flips (flipping a variable is a changing its value).

Algorithm GSAT

procedure GSAT(A,Max_Tries,Max_Flips)

A: is a CNF formula

for i:=1 to Max_Tries do

S <- instantiation of variables

for j:=1 to Max_Iter do

if A satisfiable by S then

return S

endif

V <- the variable whose flip yield the most important raise in the number of satisfied clauses;

S <- S with V flipped;

endfor

endfor

return the best instantiation found

end GSAT

Several heuristics have been implemented within GSAT in order to efficiently solve structured problems.

- Random-Walk
- The implementation of random walk within the GSAT algorithm is similar to MCRW.

- Clause weight
- This technic result from the observation that for some problems, several resolution attempts reach the same unsatisfied final set of clauses. So, each clause has not the same weight on the resolution, some clauses will be much harder to solve. The resolution process must offer more importance to these "hard" clauses.
- A way to deal with this kind of problems is to associate a weight to each clause, in order to modify its influence on the global score. Thanks to this weight heuristic, the participation of a satisfied "hard" clause is more important. Furthermore the weight can be automatically found using the following method:
    - initialize each clause weight to '1'
    - at the end of each tries, add '1' to each unsatisfied clause weight.
- Averaging in previous near solutions
- After each attempt, GSAT restarts with a random initial problem variables. This heuristic offers to reuse parts of the best assignments issued from the two previous states. Therefore, the starting variables vector for i-th attempt is computed from the bitwise* of the two best reached states during the attempts (i-2)-th and (i-1)-th.

* identical bits representing identical variable are reused, the other ones are randomly chosen.

6.10.7 Local-Search

All above algorithms are based on common idea known under the notion local search. In local search, an initial configuration (valuation of variables) is generated and the algorithm moves from the current configuration to a neighborhood configurations until a solution (decision problems) or a good solution (optimization problems) has been found or the resources available are exhausted. This idea is expressed in the following general local search algorithm that enables implementation of many particular local search algorithms via definitions of specific procedures.

Algorithm Local-Search

```
procedure local-search(Max_Moves,Max_Iter)
s <- random valuation of variables;
for i:=1 to Max_Tries while Gcondition do
for j:=1 to Max_Iter while Lcondition do
if eval(s)=0 then
return s
endif;
```

```
select n in neighborhood(s);
if acceptable(n) then
s <- n
end if
endfor
s <- restartState(s);
endfor
return s
end local-search
```

References

Alidaee, B, G. Kochenberger, and A. Ahmadian, "0-1 Quadratic Programming Approach for the Optimal Solution of Two Scheduling Problems," *International Journal of Systems Science*, 25, 401-408, 1994.

Amini, M., B. Alidaee, G. Kochenberger, "A Scatter Search Approach to Unconstrained Quadratic Binary Programs," *New Methods in Optimization*, Cone, Dorigo, and Glover, eds., McGraw-Hill, 317-330, 1999.

Borchers, B. and J. Furman, "A Two-Phase Exact Algorithm for Max-SAT and Weighted Max SAT," *J. of Combinatorial Optimization*, 2,299-306 (1999).

Boros, E., and P. Hammer, "The Max-Cut Problem and Quadratic 0-1 Optimization, Polyhedral Aspects, Relaxations and Bounds," *Annals of OR*, 33,151-225 (1991)

Boros, E. and P. Hammer, " Pseudo-Boolean Optimization," *Discrete Applied Mathematics*, 123(1-3), 155-225 (2002)

Boros, E, P. Hammer, and X, Sun, "The DDT Method for Quadratic 0-1 Minimization," RUTCOR Research Center, RRR 39-89, 1989.

Boros, E. and A. Prekopa, "Probabilistic Bounds and Algorithms for the Maximum Satisfiability Problem," *Annals of OR*, 21 (1989), pp. 109-126.

Bourjolly, J.M., P. Gill, G. Laporte, and H. Mercure, "A Quadratic 0/1 Optimization Algorithm for the Maximum Clique and Stable Set Problems," Working paper, University of Montreal, (1994).

Chardaire, P, and A. Sutter, "A Decomposition Method for Quadratic Zero-One Programming," *Management Science*, 41:4, 704-712, 1994.

Cho, J-D, S. Raje, and M. Sarrafzadeh, "Fast Approximation Algorithms on Maxcut, K-Coloring, and K-Color Ordering for VLSI Applications," IEE Transactions on Computers, 47, 1253-1266, 1998

De Simone, C. M. Diehl, M. Junger, P. Mutzel, G. Reinelt, and G. Rinaldi, 'Exact Ground State4s of Ising Spin Glasses: New Experimental Results with a Branch and Cut Algorithm," *Journal of Statistical Physics*, 80, 487-496 (1995)

Du, D. and P. Pardalos, (ed), *Handbook of Combinatorial Optimization*, (4 volumes), Kluwer Academic Publishers, 1998-99.

Gallo, G, P. Hammer, and B. Simeone, "Quadratic Knapsack Problems," *Mathematical Programming*, 12, 132-149, 1980.

Glover, F, and M. Laguna, "Tabu Search," Kluwer Academic Publishers, 1997.

Glover, F., "A Template for Scatter Search and Path Relinking," School of Business,

University of Colorado, Technical Report, February 1998.

Glover, F., B. Alidaee, C. Rego, and G. Kochenberger, "One-Pass Heuristics for Large-Scale Unconstrained Binary Quadratic Programs," *EJOR 137, pp. 272-287,* 2002.

Glover, F., G. Kochenberger, B. Alidaee, and M.M. Amini, "Tabu with Search Critical Event Memory: An Enhanced Application for Binary Quadratic Programs," In: *MetaHeuristics: Advances and Trends in Local Search Paradigms for Optimization,* (Eds.) S. Voss, S. Martello, I. Osman, and C. Roucairol. Kluwer Academic Publisher, Boston, 1999.

Glover, F., G. Kochenberger., and B. Alidaee, "Adaptive Memory Tabu Search for Binary Quadratic Programs," Management Science, 44:3, 336-345, 1998.

Grotschel, M., M. Junger, and G. Reinelt, "An Application of Combinatorial Optimization to Statistical Physics and Circuit Layout Design," *Operations Research,* Vol.36, #3, May-June, (1988), pp. 493-513.

Hammer, P., P. Hansen, and B. Simone, "Upper Planes of Quadratic 0/1 Functions and Stability in Graphs," Nonlinear Programming 4, O. Mangasarian, R. Meyer, and S. Robinson (eds), Academic Press, New York, (1981), pp. 395-414.

Hammer, P., and S. Rudeanu, *Boolean Methods in Operations Research,* Springer-Verlag, New York, 1968.

Hansen, P.B., "Methods of Nonlinear 0-1 Programming," *Annals Discrete Math,* vol. 5, pp.53-70, 1979.

Hansen, P, B. Jaumard., and V. Mathon, "Constrained Nonlinear 0-1 Programming," *INFORMS Journal on Computing*, 5:2, 97-119, 1993.

Hansen, P. and B. Jaumard, "Algorithms for the Maximum Satisfiability Problem," *Computing*, 44, 279-303 (1990).

Iasemidis, L. D., D. S. Shiau, J.C. Sackellares, and P. Pardalos, "Transition to Epileptic Seizures: Optimization," *DIMACS Series in Discrete Math and Theoretical Computer Science,* Vol. 55, (2000), pp. 55-73.

Joseph, A. "A Concurrent Processing Framework for the Set Partitioning Problem," Computers & Operations Research, 29, 1375-1391, 2002.

Kochenberger, G., F. Glover, B. Alidaee, and C. Rego, "Solving Combinatorial Optimization Problems via Reformulation and Adaptive Memory Metaheuristics,"Working Paper, University of Colorado at Denver, 2002.

Kochenberger, G., F. Glover, B. Alidaee, and C. Rego, " An Unconstrained Quadratic Binary Programming Approach to the Vertex Coloring Problem," Working Paper, University of Colorado at Denver, 2002.

Krarup, J, and A. Pruzan, "Computer Aided Layout Design," *Mathematical Programming Study*, 9, 75-94, 1978.

Laughunn, D.J, "Quadratic Binary programming," *Operations Research*, 14, 454-461, 1970.

McBride, R.D., and J.S. Yormack, "An Implicit Enumeration Algorithm for Quadratic Integer Programming," *Management Science*, 26, 282-296, 1980.

Mingozzi, A., M. Boschetti, S. Ricciardelli and L. Blanco, "A Set Partitioning Approach to the Crew Scheduling Problem," Operations Research, 47 (6) 873- 888, 1999.

Palubeckis, G. "A Heuristic-Branch and Bound Algorithm for the Unconstrained Quadratic Zero-One Programming Problem," *Computing*, pp. 284-301, (1995).

Pardalos, P, and G.P. Rodgers, "Computational Aspects of a Branch and Bound Algorithm for Quadratic Zero-one Programming," *Computing*, 45, 131-144, 1990.

Pardalos, P, and G.P. Rodgers, "A Branch and Bound Algorithm for Maximum Clique problem," *Computers & OR*, 19, 363-375, 1992.

Pardalos, P, and J. Xue, "The Maximum Clique Problem," *The Journal of Global Optimization*, 4, 301-328, 1994.

Phillips, A.T., and J.B. Rosen, "A Quadratic Assignment Formulation of the Molecular Conformation Problem," *The Journal of Global Optimization*, 4, 229-241, 1994.

Rosenberg, I.G., "0-1 Optimization and Non-linear Programming." *Revue Francaise d'Automatique, d'Informatique et de Rescherche Operationnelle,* (S\'erie Blueu) 2, 95-97, (1972).

Tsuda, N., "Fault-Tolerant Processor Arrays Using Additional Bypass Linking Allocated by Graph-Node Coloring," IEEE Transactions on Computers, 49, 431-442, 2000.

Witsgall, C., "Mathematical Methods of site Selection for Electronic System (EMS)," NBS Internal Report, 1975.

Index

www.ingramcontent.com/pod-product-compliance
Lightning Source LLC
LaVergne TN
LVHW011818200926
848484LV00005B/1234